Lecture Notes in Mathematics

Edited by A. Dold and B. Eckmann

892

Helmut Cajar

Billingsley Dimension in Probability Spaces

Springer-Verlag
Berlin Heidelberg New York 1981

Author

Helmut Cajar
Mathematisches Institut A, Universität Stuttgart
Pfaffenwaldring 57, 7000 Stuttgart 80
Federal Republic of Germany

AMS Subject Classifications (1980): 10K50, 28D99, 60F15, 60J10

ISBN 3-540-11164-6 Springer-Verlag Berlin Heidelberg New York
ISBN 0-387-11164-6 Springer-Verlag New York Heidelberg Berlin

2141/3140-543210 **D 93**

TABLE OF CONTENTS

Introduction

A number of authors (see, e.g., Besicovitch [7], Knichal [32], Eggleston [23], [24], Volkmann [44], [45], [47], [49] and Cigler [19]) have computed the Hausdorff dimensions (h-dim) of sets of real numbers characterized by digit properties of their g-adic representations. A detailed comparison of the results of these papers shows the following phenomenon: If the Hausdorff dimension of some non-denumerable union of sets M_α, $\alpha \in I$, of the type under consideration as well as the Hausdorff dimensions of the individual sets M_α are known, possibly from different sources, then the relation

$$\text{(SUP)} \quad \text{h-dim}\left(\bigcup_{\alpha\in I} M_\alpha\right) = \sup_{\alpha\in I} \text{h-dim}(M_\alpha)$$

holds, while this equality is, in general, only true for denumerable unions.

In the papers cited above each real number $r \in [0,1]$ is expressed by its g-adic expansion

$$r = \sum_{i=1}^{\infty} \frac{e_i}{g^i} = (e_1, e_2, \ldots), \ e_i \in \{0,1, \ldots, g-1\}$$

for some fixed integer $g \geqslant 2$. The study of dimensions is not affected here by the fact that this expansion is ambiguous for denumerably many r. Then the relative frequency $h_n(r, j)$ by which the digit j occurs in the finite sequence $(e_1, e_2, \ldots, e_n)$ is introduced. The sets under investigation are always of the type which can be described in terms of the limit points of the sequence $\{(h_n(r, 0), h_n(r, 1), \ldots, h_n(r, g-1))\}_{n\in\mathbb{N}}$ of g-tuples in the Euclidean space $\mathbb{R}^g$. Volkmann [49] computed the Haudorff dimension of the smallest sets which can be characterized in this fashion.

If we consider also the relative frequencies of blocks $(j_1, j_2, \ldots, j_l)$ of digits within the sequence $(e_1, e_2, \ldots)$ simultaneously for all blocks of arbitrary length l then the limit points obtained by letting n tend to infinity may be identified with special probability measures (W-measures), to be called distribution measures. Sets of real numbers which may be characterized solely by means of distribution measures shall be called saturated sets in the sequel.

Colebrook [20] computed the Hausdorff dimension of the smallest saturated sets. His results also yield the relation (SUP) for saturated sets M whenever both sides of the equation (SUP) are known.

In the present paper the relation (SUP) is proved for arbitrary saturated sets M and arbitrary sets I of indices. Furthermore we shall replace Hausdorff dimension by a more general Billingsley dimension with respect to a non-atomic, ergodic Markov measure P over a sequence space with finite state space. The key for proving the relation (SUP) lies in the representation of the Billingsley dimension as infimum of certain μ-P-dimensions which always satisfy the equation (SUP) trivially. This infimum of the μ-P-dimension shall be investigated in Chapter I within a general frame-work. It shall be studied extensively as a dimension of its own right, to be called P-dimension. In Chapter II we shall compute the Billingsley and Hausdorff dimensions of the smallest saturated sets and establish the relation (SUP) for arbitrary saturated sets, with applications in both directions.

At the beginning of Chapter I we shall state general remarks and conventions in § 1.A. A summary of the ergodic theory of a sequence space with denumerable state space, to the extent as it is needed in the present paper, will be given in § 1.B.

Billingsley [9] uses a fixed stochastic process in order to define his dimension. In § 2 this process shall be replaced by a dimension system. A dimension system consists of a basic space X on which a sequence of decompositions is defined in such a way that each of them is a refinement of the preceding one. These decompositions, in turn, generate a σ-algebra on X. In a dimension system a μ-P-dimension (μ-P-dim(M)) is defined for arbitrary subsets M of X for any given propability measure (W-measure) μ and any non-atomic W-measure P. The infimum of all μ-P-dimensions of a set M, extended over all W-measures μ, is then called the P-dimension of M, written P-dim(M), as mentioned above. The basic properties of these concepts are stated. A number of theorems which Billingsley [10,§ 2] has shown to hold for his dimension are also valid analogously for the P-dimension. In correspondence to the elementary nature of the definitions a large part of the proofs is also elementary, and a first simple criterion for the validity of the relation (SUP) for P-dimensions (Theorem 2.7) is obtained: If there exists a W-measure μ such that, for all set M_α, the μ-P-dimension is equal to the P-dimension, then the equation

$$\text{(SUP)} \qquad \text{P-dim}\Big(\bigcup_{\alpha\in I} M_\alpha\Big) = \sup_{\alpha\in I} \text{P-dim}(M_\alpha)$$

is true.

A comparison between P-dimension and Billingsley dimension relative to P is given in § 3. The P-dimension of a set is never smaller than its Billingsley dimension. However, if a dimension system satisfies a certain completeness condition which always holds for sequence spaces with denumerable state space, then both dimensions coincide (Theorem 3.3). Now it is possible to express the Hausdorff dimension as a P-dimension by means of the theorem of Wegmann [51, Satz 2], thus reaching the original problem (SUP) again. Even though one might, in the light of this section, disregard the concept of P-dimension in addition to Billingsley's dimension in many cases, it is nevertheless justified to maintain the former and to investigate it within general dimension systems on account of the elementary approach to definitions and implied properties which it provides.

In view of the criterion stated above concerning the validity of the relation (SUP) for P-dimensions it is of interest to know W-measure which are "as small as possible" or to know lower bounds for sufficiently large families of W-measures relative to the partial ordering "less or equal by dimension" on the set of all W-measures (Def. 2.6) introduced in § 2 already. For this purpose a quasimetric q on the set of all W-measures over a dimension system is introduced (Def. 4.1) and investigated in § 4.A. In addtion to results on the continuity of the μ-P-dimension and the P-dimension with respect to μ and P it is in particular shown that the family of invariant Markov measures of arbitrary order on a sequence space with finite state space is bounded from below by dimension (Theorem 4.5). This shall be of particular interest in Chapter II for establishing the relation (SUP) for saturated sets. In § 4.B we shall consider Markov kernels in order to construct lower bounds for families of W-measures in a more general (and partially more elegant) fashion.

§ 5 is not needed for Chapter II. In this section we define the P-dimension of W-measures in analogy to Kinney/Pitcher [31] who introduced a Hausdorff dimension of W-measures on the interval [0,1]. By means of §§ 2 and 4 we obtain quickly some interesting connections such as a representation of a P-dimension of invariant W-measures in terms of the P-dimension of ergodic W-measures (Theorem 5.5), which augments known representation theorems for invariant W-measures (compare, e.g., Lemma 1.2) as far as their P-dimension is concerned.

Then Chapter II deals with the Billingsley dimension (relative to a non-atomic, ergodic Markov measure P) of the saturated sets of a sequence space with finite state space. As mentioned already, Billingsley dimension may be written as P-dimension in this context such that all the tools of Chapter I are applicable.

First we deal in § 6.A with the distribution measures of a single point of the sequence space in order to be able to define exactly the saturated sets and the smallest saturated sets. Various preliminary arguments are given in § 6.B and § 6.C after which we discuss in § 6.D the problem how and to what extent the Markov measure P, assumed to be non-atomic and ergodic, may be replaced by a more general W-measure P'. § 6.E lists those functions and some of their properties which occur in the following sections.

In § 7 we first give an upper bound for the Billingsley dimension of the smallest saturated sets (Theorem 7.1). In view of the aim of establishing the relation (SUP) for saturated sets an essential role is played by a certain μ_0-P-dimension which serves as an upper bound.

In order to provide a lower bound for the Hausdorff dimension of the smallest saturated set, Colebrook [20] constructs a certain subset of it which consists of all numbers whose g-adic expansion is obtained by juxtaposition of certain specified digit blocks of growing lengths. But only those numbers contribute something to the P-dimension of the subset so constructed for which all transitions from one specified block to the next have positive P-probabilities, and thus Colebrook's procedure can not be applied here.

Therefore we construct a suitable W-measure in order to obtain a lower bound for the Billingsley dimension of the smallest saturated sets (Theorem 7.2), a procedure which is typical for the construction of Billingsley dimensions. But the W-measure which we introduce has an additional property which shall permit us in § 9.C and § 9.D to study sets characterized by the absolute non-occurrence of digits or digit blocks in addition to saturated sets. Finally we are able to compute the Billingsley dimensions of the smallest saturated sets. Now an infimum principle appears as in Colebrook [20] , but in Colebrook's paper the Hausdorff dimension of a smallest saturated set equals, up to the factor ln g, the infimum of the entropies of the distribution measures which describe the set, whereas in the study of a more general P-dimension the entropy has to be modified by a factor which depends on the distribution measure and on P. At the end of § 7 we show that even the smallest saturated sets permit non-denumerable decompositions into subsets with the same dimension.

In § 8 we determine the Billingsley dimension of an arbitrary saturated set which turns out to be equal to the supremum of the Billingsley dimensions of those smallest saturated sets which it contains. This implies immediately the relation (SUP) for the saturated sets. The results of § 3 enable us to obtain the corresponding propositions on the Hausdorff dimension of saturated subsets of the interval [0,1] with respect to g-adic expansions (Theorems 8.1 and 8.2). Thus the "maximum entropy principle" as observed by Billingsley [9] turns out to be a general "supremum infimum principle" of the Billingsley dimension of saturated sets. Furthermore this section contains examples illustrating these theorems and results on the continuity of the P-dimension with respect to the saturated set under consideration.

In § 9 we use the results at our disposal in order to describe the Billingsley or Hausdorff dimension of sets of certain types. In § 9.A we group the points of the sequence space together into saturated sets whose distribution measures are contained in, or contain a given set of W-measures or at least one W-measure from it. The sets studied in § 9.B are obtained by considering only relative frequencies of the digit blocks of some given length. By means of the results so obtained we can prove those of Besicovitch [7], Knichal [32], Eggleston [23] and [24], Volkmann [44], [45], [47], [49], Cigler [19] and Billingsley [9] and [10] as far as they are concerned with Hausdorff or Billingsley dimensions of saturated sets.

In § 9.C we consider, in addition to describing sets by distribution measures, the stipulation that finitely many blocks of digits do not occur in the sequence representing the points of the sequence space. This leads to intersections of saturated sets with Cantor-type sets. Here we extend the relation (SUP) to a class of subsets of the sequence space of which the saturated sets form a proper subclass. This enables us also to determine the Hausdorff and Billingsley dimension, respectively, of the sets studied by Volkmann [46] and [48] and by Steinfeld/Wegmann [43] by the approach taken here. Finally we determine in § 9.D the Billingsley dimension of sets which are characterized by the non-occurrence of denumerably many digit blocks.

As a whole, Chapter II and § 9, in particular, furnish results and methods by means of which the computation of the Billingsley or Hausdorff dimension of a set may be reduced in many cases to an extremal value problem with constraints which in turn may be solved, e.g., by standard methods of calculus. In this context it should be mentioned that the concepts of upper and lower noise ("bruit supêrieur", "bruit infêrieur") of a real number, as introduced by Rauzy [38] for the description of deterministic numbers, provide further examples of saturated sets. By our methods we obtain immediately a theorem of Bernay [6] according to which the set of deterministic numbers has Hausdorff dimension zero.

More general expansions of real numbers (see Galambos [26]) do not always lead to saturated sets for which the methods of Chapter II are directly applicable. Thus the results of Schweiger [40] and Schweiger/Stradner [41], investigating digit extensions of more general arithmetic transformations, are only covered in special cases by these methods. This is true, in particular, in the case of ϑ-adic digit expansions for a real number $\vartheta > 1$ with a terminating ϑ-adic representation of the number one inasmuch as, according to Cigler [18], the digits, considered as random variables on [0,1], form an ergodic Markov chain then.

With some additional effort Chapter II could be modified in order to cover Cantor series as studied by Peyrière [36]. The continued fraction algorithm, however, is too difficult to be treated by a modification of Chapter II. Nevertheless, one might expect in view of the results of Billingsley/Henningsen [13] that a theory of saturated sets should be true which is largely analogous to Chapter II. Within the scope of the present paper no attempt has been made to cover this subject. Many authors (see, e.g., R.C. Baker [2], Beardon [4], Beyer [8], Boyd [14], Hawkes [30], Nagasaka [34], Pollington [37] and others) have determined Hausdorff dimensions of sets which can not be interpreted as saturated sets. For these problems Chapter II is of little use. Perhaps parts of Chapter I might be useful in order to obtain simplified computations of the dimensions under consideration. But from the point of view taken in Chapter I saturated sets are only an example of a family of sets with the property (SUP). The paper by Baker and Schmidt [1] shows that families of sets with the property (SUP) may be obtained by means of criteria for the approximation of real numbers by algebraic numbers.

It remains to mention that the concept of P-dimension by itself is not sufficient in order to study Hausdorff measures (for the definition see Rogers [39], and for examples, see Hatano [29] and Steinfeld/Wegmann [43]). These measures may be obtained by several approaches such as metrizing the given space (see Wegmann [50]) or using the method of Sion/Willmott [42].

We should mention two directions in which the concept of P-dimension may be generalized. For one, μ-P-dimensions may be defined not only for W-measures μ but also for more general valuations μ of the cylinders which form the elements of the decompositions (compare Remark 3.2.4). In this way, for each family of valuations μ a concept of dimension is obtained as infimum of the corresponding μ-P-dimensions. Furthermore we might, instead of defining the dimension system by means of a sequence of decompositions of a given space, restrict ourselves to a subsequence, thus obtaining a new definition of dimension. This would be analogous to a generalization of the definition of Hausdorff dimension as studied by Buck [15] and [16]. We shall not go into these two kinds of generalizations, neither shall we investigate the problem under which conditions a dimension so modified coincides with the original one.

I wish to express my sincere thanks to my teacher and Ph.D. supervisor, Professor Dr. Bodo Volkmann, who suggested the subject of the present thesis and supported my work with patience. He also initiated the necessary steps for the publication of this paper as part of the Springer Lecture Notes and provided his help in the preparation of the English version of the text.

My thanks also go to my friend, Dr. Konrad Sandau, for frequent discussions from which many ideas of this paper emerged.

I am indebted to the Springer Publishing Company for having accepted this paper for the "Lecture Notes in Mathematics". Last not least, I owe thanks to Mrs. Elisabeth Schlumberger for the time consuming and strenuous effort of typing the manuscript with extreme care.

Stuttgart, July 1981 Helmut Cajar

CHAPTER I

P-dimension

§ 1. Preliminaries, notation, terminology

1.A Generalities. Definitions, theorems, lemmas, remarks and examples are numbered consecutively in each section. The number of a theorem is also designed to its corollaries. Theorems which we quote from literature are not numbered. No originality is claimed for the content of lemmas even when a proof is given. The end of each proof is marked by the symbol //. Furthermore, the following symbols are used:

$\emptyset$ for the empty set,
$\mathbb{N}$ for the set of natural numbers,
$\mathbb{N}_0 = \mathbb{N} \cup \{0\}$,
$\mathbb{R}$ for the set of real numbers,
$\mathbb{R}^+$ for the set of positive real numbers,
$\mathbb{R}_0^+$ for the set of non-negative real numbers,
λ for Lebesgue measure on $\mathbb{R}$,
N for any subset of $\mathbb{N}$,
I for any (not necessarily denumerable) set of indices,
$|A|$ for the cardinality of a denumerable set A,
χ_A for the characteristic function of a set A.

For brevity, a sequence $\{a_n\}_{n\in N}$ is also denoted by $\{a_n\}_n$ or simply by $\{a_n\}$. A denumerable set is either finite or countably infinite. Expressions like "inf" and "sup" are permitted to assume values $+\infty$ and $-\infty$. Unless stated otherwise, $\sup \emptyset = 0$ and $\inf \emptyset = +\infty$. The symbols $+\infty$ and $-\infty$ are treated by the ordinary rules of the extended real number system. Furthermore, we use the following conventions:

$$\ln 0 = -\infty, \qquad 0 \cdot \ln 0 = 0,$$

$$\frac{\ln c}{\ln c} = 1 \qquad \forall\, c \in [0,1],$$

$$\frac{\ln 0}{\ln c} = +\infty \qquad \forall\, c \in (0,1],$$

$$\frac{\ln c}{\ln 1} = +\infty \qquad \forall\, c \in [0,1),$$

where ln denotes the natural logarithm.

A metric or quasimetric is also allowed to assume the value $+\infty$. In a quasimetric space (M, δ) we denote the open ε-neighbourhood of a point $x \in M$ and of a subset H of M, respectively, by

$$U(x, \varepsilon) = U^{\delta}(x, \varepsilon) = \{y \in M \mid \delta(x, y) < \varepsilon\}$$

and by

$$U(H, \varepsilon) = U^{\delta}(H, \varepsilon) = \{y \in M \mid \exists\, x \in H : \delta(x, y) < \varepsilon\}.$$

The distance between two non-empty subsets A and B of M is denoted by

$$\delta(A, B) = \inf \{\delta(x, y) \mid x \in A, y \in B\}.$$

A metric or quasimetric δ is explicitly mentioned in topological concepts related to the topology induced by δ whenever it is not clear from the context or if it is different from the ordinary metric or quasimetric on the space considered. In this sense we use the terms δ-closed, δ-separable, or $\bar{A}^{\delta}$ for the closure of a subset A of M relative to the δ-topology.

Definitions from measure and probability theory which are not given here may be found, e.g., in Bauer [3]. A probability measure (W-measure for short) on a measurable space $(X, \underline{X})$ (where X denotes the basic set and $\underline{X}$, a σ-algebra on X) is a measure m on $(X, \underline{X})$ with $m(X) = 1$. The expression "$\forall[m]\ x \in B$" means "for m-almost all $x \in B$". The essential supremum of a non-negative measurable function f on X over a measurable set M (written $m\text{-ess.sup}_{x \in M} f(x)$) is understood to equal zero if $m(M) = 0$. The restriction of the measure m to a measurable subset B of X, defined on the σ-algebra $\{A \in \underline{X} \mid A \subset B\}$, is denoted by $m|_B$. If two measurable sets A and B are given, the conditional probability

$$m(A/B) = \frac{m(A \cap B)}{m(B)}$$

is defined only if $m(B) > 0$. Products involving undefined conditional probabilities are understood to be zero if they have at least one vanishing factor. Otherwise the product remains undefined. The set of all W-measures on a measurable space $(X, \underline{X})$ is denoted by Π. Thus Π is a convex subset of the linear space of all finite signed measures on $(X, \underline{X})$. The convex closure of a subset Θ of Π is denoted by $\langle\Theta\rangle$ and also by $\langle\mu_1, \mu_2, \ldots, \mu_n\rangle$ if $\Theta = \{\mu_1, \mu_2, \ldots, \mu_n\} \subset \Pi$ is a finite set. A convex linear combination $\Sigma_{i \in N}\alpha_i\mu_i$ $(\alpha_i \geq 0, \Sigma_{i \in N}\alpha_i = 1)$ of W-measures $\mu_i \in \Pi$ is also defined for countably infinite index sets N and is itself a W-measure.

Definition 1.1. A convex linear combination $\Sigma_{i \in N}\alpha_i\mu_i$ of W-measures μ_i is called a non-trivial convex combination if $\alpha_i > 0$ for all $i \in N$. By a face of the convex set Π we mean any convex subset Θ of Π satisfying

$$\Theta \cap \langle\mu, \nu\rangle \setminus \{\mu, \nu\} \neq \emptyset \Rightarrow \langle\mu, \nu\rangle \subset \Theta \qquad \forall\, \mu \in \Pi\ \forall\, \nu \in \Pi.$$

By the face of a W-measure $\mu \in \Pi$, to be written as $\Sigma(\mu)$, we mean the smallest face of Π which contains μ.

Remark 1.1. Since the intersection of arbitrarily many faces of Π is again a face, the face of a W-measure is well-defined. It has the following properties:

(1) $\Sigma(\mu)$ is convex.

(2) $\Sigma(\mu) = \{\nu \in \Pi \mid \exists\, \nu' \in \Pi \quad \exists\, \alpha \in (0,1) : \mu = \alpha\nu + (1 - \alpha)\nu'\}$.

(3) $\nu \in \Sigma(\mu) \Rightarrow \Sigma(\nu) \subset \Sigma(\mu)$.

(4) $\Sigma(\mu') = \Sigma(\mu) \Longleftrightarrow \exists\, \nu, \nu' \in \Pi \;\; \exists\, \alpha, \beta \in (0,1) : \mu = \alpha\nu + (1-\alpha)\nu', \; \mu' = \beta\nu + (1-\beta)\nu'$.

(5) If μ is non-atomic (i.e. $\forall\, B \in \underline{X}\; \exists\, D \in \underline{X} : D \subset B,\ \mu(D) = \frac{1}{2}\mu(B)$) then all $\nu \in \Sigma(\mu)$ are non-atomic.

(6) $\Sigma(\mu)$ is the set of all W-measure $\nu \in \Pi$ which are absolutely continuous with respect to μ and whose density $\frac{d\nu}{d\mu}$ is μ-almost certainly bounded.

1.B The sequence space $A^{\mathbb{N}}$

For an introduction to ergodic theory the reader should consult, e.g., Billingsley [11] or Denker/Grillenberger/Sigmund [21].

Let A be a denumerable set and let $X := A^{\mathbb{N}}$ be the space of all sequences in A. The set A is called the state space and its elements, states. Subsets of X of the form

$$[\underline{b}] := [b_1, b_2, \ldots, b_l] := \{(x_1, x_2, \ldots) \in X \mid (x_1, x_2, \ldots, x_l) = \underline{b}\}, \quad \underline{b} = (b_1, b_2, \ldots, b_l) \in A^l$$

are called cylinders of order l. For any block $\underline{b} = (b_1, b_2, \ldots, b_l) \in A^l$, the symbol $\underline{b}'$ denotes the block $\underline{b}' := (b_1, b_2, \ldots, b_{l-1}) \in A^l$. For any point $x \in X$, let $Z_0(x) = X$ and let $Z_n(x)$ be the cylinder of order n containing the point x. Similarly, we let

$$[\underline{b}] := X \text{ for all } \underline{b} = (b_1, \ldots, b_l) \in A^0.$$

By $\underline{X}$ we denote the smallest σ-algebra on X containing all cylinders of all orders. The shift T is the measurable mapping of X onto itself defined by

$$T(x_1, x_2, \ldots) := (x_2, x_3, \ldots) \qquad \forall\, (x_1, x_2, \ldots) \in X.$$

A W-measure on the measurable space $(X, \underline{X})$ is uniquely determined by its values on all cylinders. On the other hand, any set function μ defined on the cylinders and satisfying the conditions

(W1) $\Sigma_{b \in A}\mu([b]) = 1$ and

(W2) $\Sigma_{b \in A}\mu([b_1, \ldots, b_l, b]) = \mu([b_1, \ldots, b_l]) \quad \forall\, (b_1, \ldots, b_l) \in A^l \;\; \forall\, l \in \mathbb{N}$,

may be extended to a W-measure μ on $(X, \underline{X})$ in a unique manner. Instead of $\mu([\underline{b}])$ or $\mu([b_1, b_2, \ldots, b_l])$ we also write $\mu(\underline{b})$ or $\mu(b_1, b_2, \ldots, b_l)$, respectively. In this sense, the symbol $\mu(\underline{b}/\underline{b}') = \frac{\mu(b_1,\ldots,b_l)}{\mu(b_1,\ldots,b_{l-1})}$ may be interpreted as the conditional probability for the transition of the system to the state b_l after having passed through the states $b_1, b_2, \ldots, b_{l-1}$.

A W-measure μ on $(X, \underline{X})$ is called invariant if it is invariant under the shift T, i.e. if

$$\mu(T^{-1}Y) = \mu(Y) \qquad \forall\ Y \in \underline{X}.$$

This definition is equivalent to the following condition:

(W3) $\Sigma_{b\in A}\ \mu(b, b_1, b_2, \ldots, b_l) = \mu(\underline{b}) \quad \forall\ \underline{b} = (b_1, b_2, \ldots, b_l) \in A^l \quad \forall\ l \in \mathbb{N}.$

An invariant W-measure μ on $(X, \underline{X})$ is called ergodic if each invariant set has measure 0 or 1, i.e. if all measurable sets Y have the property:

$$Y = T^{-1}Y \Rightarrow \mu(Y) \in \{0,1\}.$$

This is equivalent to the condition

$$\mu(Y \Delta T^{-1}Y) = 0 \Rightarrow \mu(Y) \in \{0,1\} \qquad \forall\ Y \in \underline{X},$$

where Δ denotes the symmetric difference. In addition to the set Π of all W-measures on $(X, \underline{X})$ we consider the set Π_{inv} of all invariant W-measures and the set Π_{erg} of all ergodic W-measures on $(X, \underline{X})$. Clearly $\Pi_{erg} \subset \Pi_{inv} \subset \Pi$.

The entropy $E(\mu)$ of an invariant W-measure μ is defined as

$$E(\mu) := \lim_{n\to\infty} \frac{-1}{n} \sum_{\underline{b}\in A^n} \mu(\underline{b}) \ln \mu(\underline{b}).$$

The quantity $\sum_{b\in A} \mu(b) \ln \mu(b)$ is finite if and only if $E(\mu)$ is finite, in which case one has has representation

$$E(\mu) = \lim_{n\to\infty} -\sum_{\underline{b}\in A^n} \mu(\underline{b}) \ln \mu(\underline{b}/\underline{b}').$$

The following Individual Ergodic Theorem of Birkhoff is true:

Theorem. Let μ be an invariant W-measure on $(X, \underline{X})$ and let f be a μ-integrable function on X. Then the sequence

$$f_n(x) := \frac{1}{n}\Sigma_{i=0}^{n-1}\ f(T^i x)$$

of arithmetical means converges μ-almost everywhere to a μ-integrable function $f^*(x)$, and the relations

$$\int f^* d\mu = \int f\, d\mu,$$

$$\lim_{n\to\infty} \int |f_n - f^*| d\mu = 0,$$

$$f^*(Tx) = f^*(x) \qquad \forall\ [\mu]\ x \in X$$

are true.

For an ergodic measure μ the limit function f^* is constant up to a μ-null set. If $f = \chi_B$ is the characteristic function of a measurable set B of X then $f_n(x)$ is the relative frequency of the points $x, Tx, T^2x, \ldots, T^{n-1}x$ contained in B. For an ergodic measure μ one then has

$$\lim_{n\to\infty} f_n(x) \quad = \quad \mu(B) \qquad\qquad \forall\ [\mu]\ x \in X.$$

<u>Shannon-McMillan-Breiman Theorem.</u> Let $\mu \in \Pi_{erg}$ with $E(\mu) < \infty$. Then

$$\lim_{n\to\infty} \frac{-1}{n} \ln \mu(Z_n(x)) \quad = \quad E(\mu) \qquad\qquad \forall\ [\mu]\ x \in X.$$

A Bernoulli measure on $(X, \underline{X})$is defined as an invariant W-measure on $(X, \underline{X})$ relative to which T is a Bernoulli shift. In this case one has

(W4) $\mu(\underline{b}) = \Pi_{i=1}^{l}\ \mu(b_i) \quad \forall\ \underline{b} = (b_1, \ldots, b_l) \in A^l \quad \forall\ l \in \mathbb{N}.$

The entropy of a Bernoulli measure is equal to

$$E(\mu) = -\Sigma_{b\in A}\ \mu(b) \ln \mu(b).$$

By Markov measures of order l, $l \in \mathbb{N}$, we mean invariant W-measures on $(X, \underline{X})$ relative to which T is a Markov shift. (In the literature, Markov measures are generally not assumed to be invariant). Markov measures of order l are characterized by the following "Markov property":

$$\mu(b_1, \ldots, b_n) = \mu(b_1, \ldots, b_{n-1}) \frac{\mu(b_{n-l},\ldots,b_n)}{\mu(b_{n-l},\ldots,b_{n-1})} \quad \forall\ (b_1, \ldots, b_n) \in A^n$$

or

$$\mu([b_1, \ldots, b_n] / [b_1, \ldots, b_{n-1}]) = \mu([b_{n-l}, \ldots, b_n] / [b_{n-l}, \ldots, b_{n-1}]) \quad \forall\, n > l.$$

The conditional probability for the transition from the state b_{n-1} to the state b_n depends only on the l states $b_{n-l}, b_{n-l+1}, \ldots, b_{n-1}$ but not on additional states of the more distant past. According to our conventions we may consider Bernoulli measures as Markov measures of order 0. The entropy of a Markov measure μ of order l is

$$E(\mu) = -\Sigma_{\underline{b}\in A^{l+1}}\ \mu(b) \ln \mu(\underline{b}/\underline{b}'), \text{ provided } -\Sigma_{b\in A}\ \mu(b) \ln \mu(b) < \infty.$$

In all other cases, $E(\mu) = \infty$. The Markov property implies that

(W5) $\mu(b_1, \ldots, b_{n+l}) = \mu(b_1, \ldots, b_l) \cdot \Pi_{i=1}^{n} \dfrac{(b_i,\ldots,b_{i+l})}{(b_i,\ldots,b_{i+l-1})}$

for all Markov measures of order l, where the right-hand side is understood to mean zero if one of the conditional probabilities involved is undefined; in this case one of the preceding factors is zero already. A Markov measure of order l is also uniquely determined by its values on the cylinders of the order l + 1.

In order to have a succinct representation for the value assumed by a Markov measure on a given cylinder, we agree to use the following two functions. For two W-measures μ and P and $l \in \mathbb{N}$, let

$$E^l(\mu, P) := -\Sigma_{\underline{b}\in A^l}\ \mu(\underline{b}) \ln P(\underline{b}/\underline{b}'),$$

where $E^l(\mu, P) = \infty$ if there exists a $\underline{b} \in A^l$ with $\mu(\underline{b}) > 0$ and $P(\underline{b}) = 0$. The function $E^l(\mu, P)$ is affine with respect to μ as long as it remains finite, i.e., for arbitrary $P, \mu, \nu \in \Pi$ with $E^l(\mu, P) < \infty$ and $E^l(\nu, P) < \infty$ the function satisfies the equation

$$E^l(\alpha\mu + (1-\alpha)\nu, P) = \alpha E^l(\mu, P) + (1-\alpha)E^l(\nu, P) \qquad \forall\, \alpha \in [0,1].$$

For any invariant W-measure μ with finite entropy it turns out that

$$E(\mu) = \lim_{l\to\infty} E^l(\mu, \mu).$$

If P is a Markov measure of order l and if μ is an invariant W-measure then

$$E^m(\mu, P) = E^{l+1}(\mu, P) \qquad \forall\, m \geq l+1.$$

In this case we shall write $E(\mu, P)$ for $E^{l+1}(\mu, P)$.

For any point $x \in X$ a sequence $\{h_n(x)\}_{n\in\mathbb{N}}$ of W-measures on $(X, \underline{X})$ is defined by the equations

$$h_n(x)(B) := \tfrac{1}{n}\Sigma_{i=0}^{n-1}\, \chi_B(T^i x) \qquad \forall\, B \in \underline{X} \qquad \forall\, n \in \mathbb{N}.$$

For a block $\underline{b} \in A^l$ we shall also write $h_n(x, \underline{b})$ or $h_n(x; b_1, \ldots, b_l)$ instead of $h_n(x)([\underline{b}])$ or $h_n(x)([b_1, \ldots, b_l])$, respectively. Clearly, $h_n(x, \underline{b})$ is the relative frequency by which the block $(b_1, \ldots, b_l)$ occurs in the sequence $(x_1, x_2, \ldots, x_{n+l-1})$, where x stands for the infinite sequence $(x_1, x_2, x_3, \ldots)$.

With this notation, a Bernoulli measure μ satisfies the relation

(W6) $$\frac{-1}{n} \ln \mu(Z_n(x)) = -\Sigma_{b\in A}\, h_n(x, b) \ln \mu(b)$$
$$= E^1(h_n(x), \mu) \qquad \forall\, x \in X \quad \forall\, n \in \mathbb{N}.$$

For a Markov measure μ of order l one has

(W7) $$\frac{-1}{n} \ln \mu(Z_n(x)) = \frac{-1}{n} \ln \mu(Z_l(x)) - \sum_{\underline{b}\in A^{l+1}} h_n(x, b) \ln \mu(\underline{b}/b')$$
$$= \frac{-1}{n} \ln \mu(Z_l(x)) + E^{l+1}(h_n(x), \mu) \ \forall\, x \in X\ \forall\, n \in \mathbb{N}.$$

Every ergodic measure μ satisfies the condition

(W8) $$\lim_{n\to\infty} h_n(x, b) = \mu(\underline{b}) \quad \forall\, \underline{b} \in A^l \quad \forall\, l \in \mathbb{N} \qquad \forall\, [\mu]\ x \in X.$$

If the state base A is finite then (W7) and (W8) imply the relation

(W9) $$\lim_{n\to\infty} \frac{-1}{n} \ln P(Z_n(x)) = E(\mu, P) \qquad \forall\, [\mu]\ x \in X$$

for any ergodic W-measure μ and any Markov measure P.

The weak topology on the set Π of all W-measures on the sequence space $(X, \underline{X})$ is the roughest topology on Π relative to which, for any $l \in \mathbb{N}$ and each block $\underline{b} \in A^l$, the mapping

$$\hat{\underline{b}}: \Pi \to \mathbb{R}; \qquad \hat{\underline{b}}(\mu) := \mu(\underline{b}) \qquad \forall\, \mu \in \Pi$$

is continuous (cf. Billingsley [12], Theorem 2.2). In this topology, a sequence $\{\mu_i\}$ converges to a W-measure $\mu \in \Pi$, if and only if the equation

$$\lim_{l\to\infty} \mu_i(\underline{b}) = \mu(\underline{b}) \qquad \forall\, \underline{b} \in A^l \qquad \forall\, l \in \mathbb{N}$$

is satisfied. For an ergodic measure μ, the equation (W8) implies

$$\lim_{n\to\infty} h_n(x) = \mu \qquad \forall\, [\mu]\; x \in X.$$

More generally, the following lemma is true:

Lemma 1.1. (Cf. Denker/Grillenberger/Sigmund [21], Prop. 5.12). Let the state space A be finite, then for any $\mu \in \Pi_{inv}$, the sequence $\{h_n(x)\}$ converges, for μ-almost all $x \in X$, to an invariant W-measure $\hat{\mu}(x) \in \Pi_{inv}$.

A given invariant W-measure μ determines, for μ-almost all $x \in X$, a mapping $x \to \hat{\mu}(x)$ of X into Π_{inv}. This mapping induces a W-measure $\tilde{\mu}$ on the σ-algebra of Borel sets Π_{inv} (the σ-algebra generated by the open sets) which has following representation properties:

Lemma 1.2. (Cf. Denker/Grillenberger/Sigmund [21], § 13). With the notation introduced above, the following assertions hold:

a) $\nu \in \Pi_{erg} \qquad \forall\, [\tilde{\mu}]\; \nu \in \Pi_{inv}$,

b) for any set $B \in \underline{X}$ one has $\mu(B) = \int \nu(B)\, d\tilde{\mu}$,

c) $E(\mu) = \int E(\nu)\, d\tilde{\mu}$,

d) $[\lim_{n\to\infty} \frac{-1}{n} \ln \mu(Z_n(x)) = E(\nu) \qquad \forall\, [\nu]\; x \in X] \qquad \forall\, [\tilde{\mu}]\; \nu \in \Pi_{inv}$.

§ 2. The P-dimension of sets

In the paper [9] and, more generally, in [10], P. Billingsley introduced a concept of dimension in probability spaces by an approach which is analogous to the introduction of Hausdorff dimension in metric spaces.

Suppose a discrete stochastic process $\{f_1, f_2, \ldots\}$ is defined on the probability space $(X, \underline{X}, P)$. The functions f_i are mappings of X into a denumerable state space A such that, for every $n \in \mathbb{N}$, the cylinders of order n, i. e. the sets of the form

$$\{x \in X \mid f_1(x) = a_1,\, f_2(x) = a_2,\, \ldots,\, f_n(x) = a_n\}\;, \qquad a_1, \ldots, a_n \in A,$$

are measurable subsets of X. A P-δ-covering of a subset M of X is a denumerable covering $\{B_i\}_{i\in N}$ of M consisting of cylinders B_i, $P(B_i) < \delta$. For any $M \subset X$, $\delta > 0$ and $\alpha > 0$ we let

$$L_P(M, \alpha, \delta) := \inf \{\Sigma_{i\in N}\, P(B_i) \mid \{B_i\}_{i\in N} \text{ is a P-}\delta\text{-covering of M}\}$$

and

$$L_P(M, \alpha) := \lim_{\delta \to 0} L_P(M, \alpha, \delta).$$

Since, for fixed M and α, the function $L_P(M, \alpha, \delta)$ is monotonically increasing for monotonically decreasing δ, the limit $L_P(M, \alpha)$ exists but it may be infinite. It is easy to see that the function $L_P(M, \alpha)$ is monotonically decreasing in α, and there exists at most one point α_0 at which it assumes a positive, finite value. Billingsley then defines

$$\text{b-dim}_P(M) := \sup \{\alpha \mid 0 < \alpha \leq 1,\ L_P(M, \alpha) = \infty\}.$$

Meanwhile it has become customary in the literature to call $\text{b-dim}_P(M)$ the Billingsley dimension of the set M, as we shall also do in the sequel. Furthermore, Billingsley proves the following theorem, primarily for the sake of completeness:

Theorem. (Billingsley [10], Th. 2.1). For any $n \in \mathbb{N}$ and $x \in X$ let $Z_n(x)$ denote the cylinder of order n containing the point x. Let μ be another W-measure on $(X, \underline{X})$, let $\delta > 0$ and assume that

$$M \subset \{x \in X \mid \liminf_{n\to\infty} \frac{\ln \mu(Z_n(x))}{\ln P(Z_n(x))} \leq \delta\}.$$

Then $\text{b-dim}_P(M) \leq \delta$. (For conventions concerning $\frac{\ln 0}{\ln 0}$, $\frac{\ln 1}{\ln 1}$ etc., see § 1.A).

The problem arises to what extent this theorem may be reversed, i.e. whether, for any set M, there exists a W-measure μ such that

$$\text{b-dim}_P(M) \geq \liminf_{n\to\infty} \frac{\ln \mu(Z_n(x))}{\ln P(Z_n(x))} \qquad \forall\, x \in M,$$

and to what extent different sets M possess the same "optimal" measure μ. These problems have motivated our introduction and investigation of a μ-P-dimension of sets and of a new concept of dimension, called P-dimension.

As Billingsley is doing in his framework, we shall only define P-dimension relative to a non-atomic W-measure P. The stochastic process $\{f_1, f_2, \dots\}$ looses its importance inasmuch as only the structure of the cylinders and decompositions determined by it are considered. In reality, Billingsley's dimension also depends only on this structure and on the W-measure P.

Definition 2.1. A dimension system $(X, \{Z_n\}_{n\in\mathbb{N}})$ is a non-empty set X together with a sequence $\{Z_n\}_{n\in\mathbb{N}}$ of denumerable decompositions of X, where $Z_0 := \{X\}$ and each Z_n is a refinement of Z_{n-1} $(n = 1, 2, \dots)$.

The cylinders of order n, $n \in \mathbb{N}_0$, are understood to be the elements of the decomposition Z_n. For any point $x \in X$ and any non-negative integer n, let $Z_n(x)$ denote the cylinder of order n which contains the point x. Furthermore, let

$\underline{Z} := \{Z_n(x) \mid x \in X,\ n \in \mathbb{N}_0\}$
be the set of all cylinders, and let $\underline{X}$ be the smallest σ-algebra containing $\underline{Z}$. Let Π denote the set of all W-measures on the measurable space $(X, \underline{X})$ and let Π_{na} denote the set of all non-atomic measures belonging to Π.

A cylinder $B \in \underline{Z}$ with $\mu(B) = 0$ for some $\mu \in \Pi$ is called a μ-null-cylinder. Two W-measures $\mu, \nu \in \Pi$ with the same null-cylinders are called weakly equivalent, written as $\mu \overset{*}{\sim} \nu$.

<u>Remark 2.1.1.</u> The cylinders are non-empty, measurable subsets of X. For any $n \in \mathbb{N}_0$ there are countably many cylinders of order n. They are mutually disjoint and they cover the space X since they form the decomposition Z_n. Thus the cylinder $Z_n(x)$ is uniquely determined for each point $x \in X$. Any cylinder of order $n + 1$ is contained in exactly one cylinder of order n.

<u>Remark 2.1.2.</u> For each point $x \in X$, the cylinders $Z_n(x)$ form a monotonically decreasing sequence of sets. A W-measure $\mu \in \Pi$ is non-atomic if and only if

$$\lim_{n\to\infty} \mu(Z_n(x)) = 0 \qquad \forall\, x \in X.$$

<u>Example 2.1.</u> Let A be a non-empty, denumerable set and let $X = A^{\mathbb{N}}$ be the sequence space on A. For any $n \in \mathbb{N}$ let the cylinders of order n be defined as the sets

$$\{(x_1, x_2, \ldots) \in X \mid (x_1, x_2, \ldots, x_n) = \underline{b}\}, \quad \underline{b} \in A^n.$$

They determine a decomposition Z_n of the space X. Thus the pair $(X, \{Z_n\})$ is a dimension system. Occasionally it is convenient to identify the decomposition Z_n with A^n. In the sequel we consider a fixed dimension system $(X, \{Z_n\})$ to be given.

<u>Definition 2.2.</u> For any two W-measures $\mu, \nu \in \Pi$ and for all points $x \in X$ we let

$$\frac{\mu}{\nu}(x) := \liminf_{n\to\infty} \frac{\ln \mu(Z_n(x))}{\ln \nu(Z_n(x))}.$$

(The conventions concerning $\frac{\ln 0}{\ln 0}$ stated in § 1.A should be borne in mind).

<u>Remark 2.2.1.</u> The symbol $\frac{\mu}{\nu}$ defines a measurable mapping of X into $\mathbb{R}_0^+ \cup \{+\infty\}$.

<u>Remark 2.2.2.</u> For any $\mu \in \Pi$, one has $\frac{\mu}{\mu} = 1$.

<u>Remark 2.2.3.</u> If $\mu(Z_n(x)) > 0 = \nu(Z_n(x))$ for all sufficiently large $n \in \mathbb{N}$ or if $\lim\limits_{n\to\infty} \mu(Z_n(x)) > 0 = \lim\limits_{n\to\infty} \nu(Z_n(x))$, then

$\frac{\mu}{\nu}(x) = 0$ and $\frac{\nu}{\mu}(x) = \infty$.

<u>Remark 2.2.4.</u> If there exists an index $n \in \mathbb{N}$ such that $\mu(Z_n(x)) = \nu(Z_n(x)) = 0$, then $\frac{\mu}{\nu}(x) = \frac{\nu}{\mu}(x) = 1$.

The following "cancellation law" is fundamental for comparing the "quotients" $\frac{\mu}{\nu}(x)$ relative to different W-measures.

Theorem 2.1. For any three W-measures μ, μ', $\mu'' \in \Pi$ the inequality

$$\frac{\mu}{\mu'} \cdot \frac{\mu'}{\mu''} \leqslant \frac{\mu}{\mu''}$$

holds at every point x, provided that the convention $0\cdot\infty = \infty\cdot 0 = 0$ is used.

Proof. For all $x \in X$ and all $n \in \mathbb{N}$ one has, using the relevant conventions,

$$\frac{\ln \mu(Z_n(x))}{\ln \mu'(Z_n(x))} \cdot \frac{\ln \mu'(Z_n(x))}{\ln \mu''(Z_n(x))} \leqslant \frac{\ln \mu(Z_n(x))}{\ln \mu''(Z_n(x))} .$$

Letting n tend to infinity, this proves the assertion.//

The second fundamental rule for handling the "quotient" $\frac{\mu}{\nu}$ is contained in the following theorem in which non-trivial convex combinations of two W-measures are considered (compare Def. 1.1).

Theorem 2.2. Let $P \in \Pi_{na}$ be a non-atomic W-measure and let $\mu = \alpha\mu' + (1 - \alpha)\,\mu''$ be a non-trivial convex combination of two W-measures μ', $\mu'' \in \Pi$. Then the relation

$$\frac{\mu}{P} = \min\,\{\frac{\mu'}{P}\,,\,\frac{\mu''}{P}\}$$

is true at each point x.

Proof. For any $x \in X$ and $n \in \mathbb{N}$ one has

$$\alpha\mu'(Z_n(x)) \leqslant \mu(Z_n(x)) \leqslant \max\,\{\mu'(Z_n(x)),\,\mu''(Z_n(x))\}.$$

Taking logarithms of these quantities and dividing by $\ln P(Z_n(x))$ we obtain

$$\frac{\ln \mu'(Z_n(x))}{\ln P(Z_n(x))} + \frac{\ln \alpha}{\ln P(Z_n(x))} \geqslant \frac{\ln \mu(Z_n(x))}{\ln P(Z_n(x))} \geqslant$$

$$\geqslant \min\,\{\frac{\ln \mu'(Z_n(x))}{\ln P(Z_n(x))}\,,\,\frac{\ln \mu''(Z_n(x))}{\ln P(Z_n(x))}\}.$$

Since $\alpha > 0$ and $P \in \Pi_{na}$, we have

$$\lim_{n\to\infty} \frac{\ln \alpha}{\ln P(Z_n(x))} = 0.$$

Now considering the lower limits of these quantities, we have

$$\frac{\mu'}{P}(x) \geqslant \frac{\mu}{P}(x) \geqslant \min\,\{\frac{\mu'}{P}(x),\,\frac{\mu''}{P}(x)\}.$$

The proof is now completed by interchanging μ' and μ''.//

Corollary 2.2.1. The theorem just proved may be extended to finite non-trivial convex combinations of W-measures.

Corollary 2.2.2. For any $P \in \Pi_{na}$ and for any denumerable non-trivial convex combination $\mu = \Sigma_{i\,N}\, \alpha_i\mu_i$ of W-measures $\mu_i \in \Pi$ the inequality

$$\frac{\mu}{P} \leqslant \inf_{i\in N} \frac{\mu_i}{P}$$

holds at all points $x \in X$.

Corollary 2.2.3. For any $\mu \in \Pi$ and any $\nu \in \Sigma(\mu)$ (see Def. 1.1 of the face $\Sigma(\mu)$) one has

$$\frac{\mu}{P} \leq \frac{\nu}{P} \qquad \forall\, P \in \Pi_{na}.$$

Corollary 2.2.4. For arbitrary $\mu, \nu \in \Pi$ with $\Sigma(\mu) = \Sigma(\nu)$ one has

$$\frac{\mu}{P} = \frac{\nu}{P} \qquad \forall\, P \in \Pi_{na}.$$

The following example shows that the symbol "$\leq$" can not be replaced by "=" in Corollary 2.2.2.

Example 2.2. Let $X := \{0,1\}^{\mathbb{N}}$ be the set of 0-1-sequences and $\{Z_n\}$ be the sequence of decompositions of X introduced in Example 1.1. By letting

$$P(B) := 2^{-n} \qquad \forall\, B \in Z_n \qquad \forall\, n \in \mathbb{N}$$

a non-atomic W-measure P on $(X,\underline{X})$ is determined. For every $i \in \mathbb{N}$ let

$x^i := (0, 0, \ldots, 0, 1, 1, 1, \ldots) \in X$, with $i - 1$ zeros preceding the ones,

and assume the measure $\mu_i \in \Pi$ to be such that

$$\mu_i(Z_i(x^i)) = 1$$

(e.g., μ_i may be atomic with $\mu_i(x^i) = 1$). Then the zero sequence $x^0 := (0, 0, 0, \ldots)$ satisfies

$$\mu_i(Z_n(x^0)) = \begin{cases} \mu_i(Z_n(x^i)) = 1, & \text{if } n < i \\ 0, & \text{if } n \geq i\,. \end{cases}$$

Hence, according to Remark 2.2.3 we have

$$\frac{\mu_i}{P}(x^0) = \infty \qquad \forall\, i \in \mathbb{N}.$$

Now we choose a number $q \in (0,1)$ and we let $\alpha_1 := 1 - q$ and $\alpha_n := q^{n-1} - q^n$. $n = 2, 3, \ldots$. For the non-trivial convex combination

$$\mu_0 := \Sigma_{i\in\mathbb{N}}\, \alpha_i \mu_i$$

this yields the relations

$$\mu_0(Z_n(x)) = \Sigma_{i>n}\, \alpha_i = q^n$$

and

$$\frac{\mu_0}{P}(x^0) = \frac{-\ln q}{\ln 2} < \inf_{i\in\mathbb{N}} \frac{\mu_i}{P}(x^0) = \infty.$$

Definition 2.3. For any $\mu \in \Pi$ and $P \in \Pi_{na}$ the μ-P-dimension of any subset $M \in X$ is defined as

$$\mu\text{-P-dim}(M) := \sup_{x\in M} \frac{\mu}{P}(x) \qquad \forall\, M \subset X.$$

Remark 2.3. The symbol μ-P-dim is a mapping of the power set of X into $\mathbb{R}_0^+ \cup \{\infty\}$. Clearly, $\mu\text{-P-dim}(\emptyset) = 0$ and $\text{P-P-dim}(M) = 1$ for any non-empty subset M of X.

The following theorem is a consequence of elementary properties of the supremum:

Theorem 2.3. Let $\mu \in \Pi$ and $P \in \Pi_{na}$. Then the following assertions are true:

a) μ-P-dim(M) is monotonic in M, i.e.,

$M_1 \subset M_2 \subset X \Rightarrow \mu\text{-P-dim}(M_1) \leqslant \mu\text{-P-dim}(M_2)$.

b) Let I be any (not necessarily denumerable) set of indices and let

$M_\alpha \subset X \quad \forall\, \alpha \in I$.

Then the relation

(SUP) $\mu\text{-P-dim}(\bigcup_{\alpha \in I} M_\alpha) = \sup_{\alpha \in I} \mu\text{-P-dim}(M_\alpha)$

holds.

The following proposition on the μ-P-dimension of sets is implied by Corollary 2.2.3:

Theorem 2.4. For any $\mu \in \Pi$ and $\nu \in \Sigma(\mu)$ (see Def. 1.1) one has

$\mu\text{-P-dim} \leqslant \nu\text{-P-dim} \qquad \forall\, P \in \Pi_{na}$.

Corollary 2.4.1. For any denumerable non-trivial convex combination $\mu_i = \Sigma_{i \in N}\, \alpha_i \mu_i$ of W-measures $\mu_i \in \Pi$ the inequality

$\mu\text{-P-dim} \leqslant \inf_{i \in N} \mu_i\text{-P-dim} \qquad \forall\, P \in \Pi_{na}$

holds.

Corollary 2.4.2. For any two W-measures $\mu, \nu \in \Pi$ with $\Sigma(\mu) = \Sigma(\nu)$ the equation

$\mu\text{-P-dim} = \nu\text{-P-dim} \qquad \forall\, P \in \Pi_{na}$

is true.

The following proposition is an immediate consequence of Theorem 2.1.

Theorem 2.5. Let $\mu, \nu \in \Pi$ and $P, P' \in \Pi_{na}$. Then for any subset M of X, letting $0 \cdot \infty = \infty \cdot 0 = 0$,

a) $\mu\text{-P-dim}(M) \geqslant \nu\text{-P-dim}(M) \cdot \inf_{x \in M} \frac{\mu}{\nu}(x)$.

b) $\mu\text{-P-dim}(M) \geqslant \mu\text{-P}'\text{-dim}(M) \cdot \inf_{x \in M} \frac{P'}{P}(x)$.

After these preparations we can now explain how each non-atomic W-measure on the dimension system $(X, \{Z_n\})$ defines a dimension.

Definition 2.4. For any non-atomic W-measure P on $(X, \underline{X})$ the P-dimension of subsets of X is defined as

$\text{P-dim}(M) := \inf_{\mu \in \Pi} \mu\text{-P-dim}(M) \qquad \forall\, M \subset X$.

In the sequel we consider P to be an arbitrary, but fixed non-atomic W-measure on $(X, \underline{X})$. The following theorem summarizes those properties of the P-dimension which justify the name "dimension" for this new concept.

Theorem 2.6. a) The inequality $0 \leqslant \text{P-dim}(M) \leqslant 1 \quad \forall\, M \subset X$ holds. For any denumerable subset M of X we have P-dim(M) = 0. The union of all P-null-cylinders has P-dimension zero, i.e.,

$$\text{P-dim}(\{x \in X \mid \exists\, n \in \mathbb{N}: P(Z_n(x)) = 0\}) = 0.$$

b) P-dim(M) is a monotonic function of M, i.e.,

$$M_1 \subset M_2 \subset X \Rightarrow 0 \leq \text{P-dim}(M_1) \leq \text{P-dim}(M_2) \leq 1.$$

c) For any set $M \subset X$ there exists a W-measure $\mu \in \Pi$ such that

$$\text{P-dim}(M) = \mu\text{-P-dim}(M).$$

d) For arbitrary, denumerably many sets $M_i \subset X$, $i \in N$ one has

$$\text{P-dim}(\bigcup_{i \in I} M_i) = \sup_{i \in I} \text{P-dim}(M_i).$$

Proof. a) Since $0 \leq \mu\text{-P-dim}(M)$ for all $\mu \in \Pi$ and $\text{P-P-dim}(M) \leq 1$ for any subset M of X (compare Remark 2.3), it follows that $0 \leq \text{P-dim}(M) \leq 1$ for all subsets M of X. Furthermore, $\text{P-dim}(\emptyset) = 0$ since we also have $\text{P-P-dim}(\emptyset) = 0$. If $M = \{x_i \mid i \in N\}$ is a non-empty, denumerable subset of X, a W-measure μ on $(X, \underline{X})$ with $\frac{\mu}{P}(x_i) = 0$ for all $x_i \in M$ is defined by the equation

$$\mu(B) := (\Sigma_{i \in N}\, 2^{-i})^{-1} \cdot \Sigma_{\substack{i \in N\\ x_i \in B}}\, 2^{-i} \qquad \forall\, B \in \underline{X}.$$

Thus $\mu\text{-P-dim}(M) = 0$ and hence $\text{P-dim}(M) = 0$. Since any decomposition Z_n is denumerable, there is a denumerable set $M = \{x_i \mid i \in N\}$ such that each cylinder $B \in \underline{Z}$ contains at least one point from the set M. With the measure μ constructed above we then have $\mu(Z_n(x)) > 0$ for all $n \in \mathbb{N}$ and all $x \in X$, and therefore $\frac{\mu}{P}(x) = 0$ for each point $x \in X$ contained in a P-null-cylinder. In view of the relation $0 \leq \text{P-dim} \leq \mu\text{-P-dim}$ this implies the last assertion of part a).

b) The monotonicity of the μ-P-dimension which we have established in Theorem 2.3 remains valid if the infimum over all $\mu \in \Pi$ is formed.

c) Let $M \subset X$ and $\{\mu_i\}_{i \in \mathbb{N}}$ be a sequence of W-measures in Π satisfying

$$\text{P-dim}(M) = \inf_{i \in \mathbb{N}} \mu_i\text{-P-dim}(M).$$

The non-trivial convex combination $\mu := \Sigma_{i \in \mathbb{N}}\, 2^{-i} \mu_i$ then satisfies the relation

$$\text{P-dim}(M) \leq \mu\text{-P-dim}(M) \leq \inf_{i \in \mathbb{N}}\ \mu_i\text{-P-dim}(M) = \text{P-dim}(M)$$

according to Corollary 2.4.1. Hence the W-measure μ satisfies the assertion c).

d) For any set $M_i \subset X$ and a denumerable family of sets $\{M_i\}_{i \in N}$ there is, according to part c) which has already been established, a W-measure $\mu_i \in \Pi$ such that $\text{P-dim}(M_i) = \mu_i\text{-P-dim}(M_i)$ (in the case $N = \emptyset$ this assertion is trivial). Therefore the W-measure

$$\mu := (\Sigma_{i \in N}\, 2^{-i})^{-1} \cdot \Sigma_{i \in N}\, 2^{-i} \mu_i$$

satisfies the following chain of inequalities in view of Theorem 2.3.b) and Corollary 2.4.1, using monotonicity of the P-dimension:

$$\text{P-dim}(\cup_{i \in N} M_i) \leq \mu\text{-P-dim}(\cup_{i \in N} M_i) = \sup_{i \in N} \mu\text{-P-dim}(M_i) \leq$$

$$\leq \sup_{i\in N} \mu_i\text{-P-dim}(M_i) = \sup_{i\in N} \text{P-dim}(M_i)$$

$$\leq \text{P-dim}(\cup_{i\in N} M_i).$$

Consequently, equality must hold everywhere in this relation, which proves part d) of the theorem.//

The following theorem establishes a condition under which Theorem 2.6.d) is also valid for non-denumerable families of sets. This case shall be of special significance in Chapter II, particularly in § 8.

Theorem 2.7. Let $\mu \in \Pi$ and I be a set of indices, not necessarily denumerable. For each $\alpha \in I$ let $M_\alpha \subset X$ and

$$\text{P-dim}(M_\alpha) = \mu\text{-P-dim}(M_\alpha).$$

Then the equation

$$\text{(SUP)} \quad \text{P-dim}\Big(\bigcup_{\alpha\in I} M_\alpha\Big) = \sup_{\alpha\in I} \text{P-dim}(M_\alpha),$$

holds and we have again

$$\text{P-dim}\Big(\bigcup_{\alpha\in I} M_\alpha\Big) = \mu\text{-P-dim}\Big(\bigcup_{\alpha\in I} M_\alpha\Big).$$

Proof. The proof of Theorem 2.6.d) is valid without changes, disregarding the construction of the W-measure μ.//

Example 2.3. Let P and β be two Bernoulli measures on the sequence space $X = A^{\mathbb{N}}$, $2 \leq |A| < \infty$, satisfying $P(b) > 0$ for all $b \in A$, and let M be the set of all sequences $x = (x_1, x_2, \ldots) \in A^{\mathbb{N}}$ in which each state $b \in A$ occurs with a relative frequency which converges to $\beta(b)$. In the terminology of § 1.B. this means that

$$M := \{x \in X \mid \lim_{n\to\infty} h_n(x, b) = \beta(b) \quad \forall\, b \in A\}.$$

For each Bernoulli measure μ on X satisfying $\mu(b) > 0$ for all $b \in A$ and for each point $x \in M$ we have (compare § 1.B, (W6)) the relation

$$\lim_{n\to\infty} \frac{1}{n} \ln \mu(Z_n(x)) = \Sigma_{b\in A}\, \beta(b) \ln \mu(b) = -E(\beta, \mu).$$

Since P is also a Bernoulli measure with the properties of the measure μ, this implies

$$\frac{\mu}{P}(x) = \frac{E(\beta,\mu)}{E(\beta,P)} \qquad \forall\, x \in M$$

and furthermore,

$$\mu\text{-P-dim}(M) = \frac{E(\beta,\mu)}{E(\beta,P)}$$

for all admissible Bernoulli measures μ. The infimum of the μ-P-dimensions, extended over all these measures, yields the relation

$$\text{P-dim}(M) \leq \frac{E(\beta)}{E(\beta,P)}\,.$$

We shall show in Example 2.4 that this upper bound for P-dim(M) is also a lower bound.

The following three theorems have been proved by Billingsley [10] with regard to his dimension. We shall formulate and prove them for P-dimension. Theorem 2.1. of Billingsley [10] as quoted at the beginning of this section, is valid for P-dimensions in view of Definitions 2.3 and 2.4. It then reads as follows:

Theorem 2.8. Let $\mu \in \Pi$, $\delta > 0$ and $M \subset \{x \in X \mid \frac{\mu}{P}(x) \leq \delta\}$. Then

$$\text{P-dim}(M) \leq \delta.$$

Proof. In short, the theorem states that

$$\text{P-dim} \leq \mu\text{-P-dim} \qquad \forall\, \mu \in \Pi,$$

which follows immediately from Definition 2.4.//

The following theorem restates Theorem 2.2 of Billingsley [10] for P-dimensions. It is obtained from Theorem 2.5 by considering the infimum relative to all W-measures $\mu \in \Pi$.

Theorem 2.9. In addition to P let P' be a non-atomic W-measure on $(X, \underline{X})$ and let M be a subset of X. Then the following inequality holds:

$$\text{P-dim}(M) \geq \text{P}'\text{-dim}(M) \cdot \inf_{x \in M} \frac{P'}{P}(x).$$

By applying Theorem 2.9 twice, interchanging the roles of P and P', we obtain Theorem 2.4 of Billingsley [10] for P-dimensions:

Theorem 2.10. In addition to P let P' be a non-atomic W-measure on $(X, \underline{X})$; furthermore let $\delta > 0$ and let M be a subset of X satisfying

$$\frac{P'}{P}(x) = \left[\frac{P}{P'}(x)\right]^{-1} = \delta \qquad \forall\, x \in M.$$

Then the following equation holds:

$$\text{P-dim}(M) = \delta \cdot \text{P}'\text{-dim}(M).$$

Remark 2.4. The assumption of Theorem 2.10 is equivalent to the statement

$$\lim_{n \to \infty} \frac{\ln P'(Z_n(x))}{\ln P(Z_n(x))} = \delta \qquad \forall\, x \in M.$$

The theorems stated so far are not sufficient for a practical computation of the P-dimension of a set M. They only furnish an upper bound for P-dim(M) if $\frac{\mu}{P}(x)$ is known for all $x \in M$ and hence μ-P-dim(M) for some W-measures μ. The following three theorems now open the possibility to establish also lower bounds for the P-dimension of a set provided the "quotient" $\frac{\mu}{P}(x)$ is known on a set M for some W-measure μ which does not assign a too small value to this set.

At first we observe that any W-measure on a dimension system generates a scale of outer measures.

Definition 2.5. For any $\mu \in \Pi$, $\delta > 0$ and any set $M \subset X$ let

$$\mu^{\delta}(M) := \inf \{ \sum_{i \in N} \mu(A_i)^{\delta} \mid A_i \in \underline{Z},\ M \subset \bigcup_{i \in N} A_i,\ N \subset \mathbb{N} \}.$$

Remark 2.5. All cylinder coverings of the set M are considered in order to determine the quantity $\mu^{\delta}(M)$. The set function μ^{δ} is an outer measure on X. Since it was agreed in Definition 2.1 to consider the set X itself as one of the cylinders, one always has

$$0 \leq \mu^{\delta} \leq 1.$$

Theorem 2.11. For all $\delta > 0$ and all $M \subset X$ the following assertion is true:

$$P^{\delta}(M) > 0 \Rightarrow P\text{-dim}(M) \geq \delta.$$

Proof. Let $\delta > 0$ and $M \subset X$ such that $P\text{-dim}(M) < \delta$. Since $P^{\delta}(A) = 0$ for any P-null-cylinder A, we may assume without loss of generality that the set M is disjoint with all P-null-cylinders. Now there exists a number $t > 0$ and a W-measure $\mu \in \Pi$ such that

$$P\text{-dim}(M) = \mu\text{-}P\text{-dim}(M) < t < \delta.$$

Hence we have $\frac{\mu}{P}(x) < t$ for all $x \in M$. Therefore there exists, for every $\varepsilon > 0$ and every point $x \in M$, a smallest positive integer $n(x)$ such that

$$\frac{\ln \mu(Z_{n(x)}(x))}{\ln P(Z_{n(x)}(x))} < t \text{ and } P(Z_{n(x)}(x))^{\delta - t} < \varepsilon.$$

The cylinders A_i involved in this argument are mutually equal or disjoint. Furthermore, they form a covering $\{A_i \mid i \in N\} := \{Z_{n(x)}(x) \mid x \in M\}$ of the set M, and we have

$$P^{\delta}(M) \leq \Sigma_{i \in N} P(A_i)^{\delta} = \Sigma_{i \in N} P(A_i)^{\delta - t} \cdot P(A_i)^{t} < \Sigma_{i \in N} \varepsilon \mu(A_i) \leq \varepsilon.$$

Since this is true for all $\varepsilon > 0$, we have $P^{\delta}(M) = 0$, which proves the assertion.//

Corollary 2.11.1. For any measurable set $M \subset X$ the following assertion holds:

$$P(M) > 0 \Rightarrow P\text{-dim}(M) = 1.$$

Corollary 2.11.2. For any subset M of X,

$$P\text{-dim}(M) \geq \inf \{\delta > 0 \mid P^{\delta}(M) = 0\}.$$

Proof. We only need to note that, for any measurable subset M of X, we always have $P(M) = P^{1}(M)$.//

By combining Corollary 2.11.1 with Theorem 2.9 we obtain a theorem for P-dimensions which is somewhat more general than Theorem 2.5 of Billingsley [10] :

Theorem 2.12. For any W-measure $\mu \in \Pi$ and for any measurable set $M \subset X$,

$$\mu\text{-ess.sup}_{x \in M} \frac{\mu}{P}(x) \leq P\text{-dim}(M).$$

Proof. Let $t := \mu\text{-ess.sup}_{x \in M} \frac{\mu}{P}(x)$. If $t = 0$ then the assertion is trivial. If $t > 0$ then the set

$M^s := \{x \in M \mid \frac{\mu}{P}(x) > s\}$

has positive μ-measure for each $s \in (0,t)$.

The conditional probability

$\nu(C) := \mu(C/M^s) \qquad \forall\, C \in X$

now satisfies the following assertions:

(1) $\frac{\nu}{P} \geq \frac{\mu}{P}$ (since $\nu \in \Sigma(\mu)$; see Def. 1.1 and Corollary 2.2.3);

(2) ν is non-atomic (otherwise $\frac{\nu}{P}(x) = 0$ for some $x \in M^s$);

(3) $\nu(M^s) = 1$.

In view of (2) we may also consider the ν-dimension of the set M^s. According to Corollary 2.11 and assertion (3) we obtain

$\nu\text{-dim}(M^s) = 1$.

Together with Theorem 2.9 and assertion (1) this further implies that

$P\text{-dim}(M) \geq P\text{-dim}(M^s) \geq \nu\text{-dim}(M^s) \cdot s = 1 \cdot s = s$.

Hence the assertion follows since this is true for each number s from the interval $(0,t)$.//

Example 2.4. Let X be the sequence space $A^{\mathbb{N}}$ with $2 \leq |A| < \infty$, let μ be an ergodic W-measure and P a non-atomic Markov measure on X (see § 1.B). Then the following relations hold for μ-almoat all $x \in X$ (compare § 1.B, the Shannon-McMillan-Breiman Theorem and (W9)):

$$\lim_{n\to\infty} \frac{-1}{n} \ln \mu(Z_n(x)) = E(\mu);$$

$$\lim_{n\to\infty} \frac{-1}{n} \ln P(Z_n(x)) = E(\mu, P) > 0.$$

For every measurable subset $Y \subset X$ with $\mu(Y) > 0$ this implies

$$P\text{-dim}(Y) \geq \mu\text{-ess.sup}_{x \in Y} \frac{\mu}{P}(x) = \frac{E(\mu)}{E(\mu,P)}.$$

Now we are able to complete the computation of Example 2.3. Inasmuch as each Bernoulli measure is ergodic, we obtain at first, using the notation of Example 2.3, the relation $\beta(M) = 1$. In view of the argument given above this implies

$$P\text{-dim}(M) = \frac{E(\beta)}{E(\beta,P)} = \frac{\Sigma_{b\in A}\, \beta(b) \ln \beta(b)}{\Sigma_{b\in A}\, \beta(b) \ln P(b)}.$$

Every W-measure $\mu \in \Pi$ yields an upper and a lower bound for the P-dimensions of measurable sets:

$$\mu\text{-ess.sup}_{x \in M} \frac{\mu}{p}(x) \leq P\text{-dim}(M) \leq \sup_{x \in M} \frac{\mu}{p}(x) \qquad \forall M \in \underline{X}.$$

According to Corollary 2.2.2, a given non-trivial convex combination of denumerably many W-measures always yields an upper bound which is at least as good as each single W-measure involved. A sharpened version of the same principle is expressed for lower bounds by the following theorem.

Theorem 2.13. Let $\mu = \Sigma_{i \in N} \alpha_i \mu_i$ be a non-trivial convex combination of denumerably many W-measures $\mu_i \in \Pi$. Then we have, for any measurable set $M \subset X$,

$$\mu\text{-ess.sup}_{x \in M} \frac{\mu}{p}(x) = \sup_{i \in N} [\mu_i\text{-ess.sup}_{x \in M} \frac{\mu_i}{p}(x)].$$

Proof. Let $t := \mu\text{-ess.sup}_{x \in M} \frac{\mu}{p}(x)$ and $t_i := \mu_i\text{-ess.sup}_{x \in M} \frac{\mu_i}{p}(x)$ for all $i \in N$.

a) Proof of "$\geq$". Let $i \in N$ be an arbitrary index. If $t_i = 0$ then trivially $t_i \leq t$. If $t_i > 0$ then we have, for each $s \in (0,t_i)$, letting

$$M_s := \{x \in M \mid \frac{\mu}{p}(x) \leq s\},$$

the chain of inequalities

$$t_i > s \geq \mu\text{-}P\text{-dim}(M_s) \geq P\text{-dim}(M_s) \geq \mu_i\text{-ess.sup}_{x \in M_s} \frac{\mu_i}{p}(x).$$

Hence $\mu_i(M \setminus M_s) > 0$ and therefore also $\mu(M \setminus M_s) > 0$. Considering these inequalities together, we obtain

$$t \geq \mu\text{-ess.sup}_{x \in M \setminus M_s} \frac{\mu}{p}(x) \geq s.$$

Since $s \in (0,t_i)$ was arbitrary, this shows that $t \geq t_i$. Since $i \in N$ was also arbitrary, we have established the relation $t \geq \sup_{i \in N} t_i$.

b) Proof of "$\leq$". This part is trivial if $t = 0$. In the case $t > 0$ let $s \in (0,t)$ be arbitrary and furthermore, we define

$$M^s := \{x \in M \mid \frac{\mu}{p}(x) \geq s\}.$$

Then $\mu(M^s) > 0$ and therefore $\mu_j(M^s) > 0$ for some $j \in N$. But according to Corollary 2.2.2 all $x \in M^s$ satisfy

$$s \leq \frac{\mu}{p}(x) \leq \frac{\mu_j}{p}(x).$$

Thus we now have

$$\sup_{i \in N} t_i \geq t_j \geq \mu_j\text{-ess.sup}_{x \in M^s} \frac{\mu_j}{p}(x) \geq s.$$

Since s was an arbitrary number from the interval $(0,t)$, this shows that

$t \leq \sup_{i \in N} t_i$.//

Remark 2.6. Let us introduce a relation $\prec$ on the set Π of all W-measures on $(X, \underline{X})$ by defining

$$\mu \prec \nu :\Leftrightarrow \mu\text{-ess.sup}_{x \in M} \frac{\mu}{P}(x) \leq \nu\text{-ess.sup}_{x \in M} \frac{\nu}{P}(x) \quad \forall M \in \underline{X} \qquad \forall P \in \Pi_{na}$$

for $\mu, \nu \in \Pi$. In this manner, the pair $(\Pi, \prec)$ becomes an upwards directed set. If we identify W-measures μ and ν whenever they satisfy $\mu \prec \nu$ and $\nu \prec \mu$ then $(\Pi, \prec)$ becomes an upwards directed semi-lattice which is σ-complete in view of the last theorem. (Concepts from lattice theory are used as defined by Gericke [27]).

We shall also consider another relation on Π which has more meaning for the applications. This relation is obtained by comparing the upper bounds for the P-dimension furnished by two W-measures μ and ν, i.e. the μ-P-dimension and the ν-P-dimension:

Definition 2.6. Let μ and ν be two W-measures on $(X, \underline{X})$. Then μ is called less or equal to ν by dimension, written as $\mu \underset{\dim}{\leq} \nu$, if

$$\frac{\mu}{P} \leq \frac{\nu}{P} \qquad \forall P \in \Pi_{na}.$$

Two W-measures μ and ν are called equal by dimension, written as $\mu \underset{\dim}{=} \nu$, if

$$\frac{\mu}{P} = \frac{\nu}{P} \qquad \forall P \in \Pi_{na}.$$

Remark 2.7.1. The relation $\underset{\dim}{\leq}$ is a partial ordering, the corresponding equivalence relation on Π being the relation $\underset{\dim}{=}$.

Remark 2.7.2. The relation $\mu \underset{\dim}{\leq} \nu$ holds if and only if

$$\mu\text{-P-dim} \leq \nu\text{-P-dim} \qquad \forall P \in \Pi_{na},$$

i.e. if μ-P-dim furnishes a better upper bound for P-dim than ν-P-dim for all non-atomic W-measures P.

Remark 2.7.3. The relation $\mu \underset{\dim}{=} \nu$ holds if and only if

$$\mu\text{-P-dim} = \nu\text{-P-dim} \qquad \forall P \in \Pi_{na},$$

i.e. if for all non-atomic W-measures P, the measures μ and ν furnish the same upper limits for the P-dimension.

Remark 2.7.4. For arbitrary measures $\mu \in \Pi$ and $\nu \in \Sigma(\mu)$ (see Def. 1.1) one has

$$\mu \underset{\dim}{\leq} \nu$$

in view of Corollary 2.2.3.

Remark 2.7.5. If two W-measures have the same faces (see Def. 1.1) then they are equal by dimension.

Remark 2.7.6. Any denumerable subset of Π is bounded from below relative to the partial ordering $\underset{\dim}{\leqslant}$ (compare Corollary 2.2.2).

Remark 2.7.7. If $\mu \in \Pi$ and $M \subset X$ are such that P-dim(M) = μ-P-dim(M) then

$$\nu \underset{\dim}{\leqslant} \mu \Rightarrow \text{P-dim}(M) = \nu\text{-P-dim}(M) \qquad \forall\, \nu \in \Pi.$$

Remark 2.7.8. By making the transition from Π to the equivalence classes of the equivalence relation $\underset{\dim}{=}$, i.e. by identifying W-measures in $(\Pi, \underset{\dim}{\leqslant})$ whenver they are equal by dimension, we obtain a downwards directed semi-lattice in which every denumerable subset is bounded from below.

The following theorem establishes an immediate connection between the relation $\mu \underset{\dim}{\leqslant} \nu$ and the "quotient" $\frac{\nu}{\mu}$.

Theorem 2.14. For all $\mu, \nu \in \Pi$ the following assertion is true:

$$\frac{\nu}{\mu} \geqslant 1 \Rightarrow \mu \underset{\dim}{\leqslant} \nu .$$

If μ is non-atomic then the converse is also true:

$$\mu \underset{\dim}{\leqslant} \nu \Rightarrow \frac{\nu}{\mu} \geqslant 1.$$

Proof. By Theorem 2.1 we have, for arbitrary $\mu, \nu \in \Pi$, $P \in \Pi_{na}$ and $x \in X$ with $\frac{\nu}{\mu}(x) \geqslant 1$, the inequality

$$\frac{\mu}{P}(x) \leqslant \frac{\mu}{P}(x) \cdot \frac{\nu}{\mu}(x) \leqslant \frac{\nu}{P}(x).$$

This implies the first assertion. The second assertion follows if we replace the W-measure P by the non-atomic W-measure μ in the condition

$$\frac{\mu}{P} \leqslant \frac{\nu}{P} \qquad \forall\, P \in \Pi_{na}.//$$

Corollary 2.14. Two non-atomic W-measures P' and P" are equal by dimension if and only if

$$\lim_{n\to\infty} \frac{\ln P'(Z_n(x))}{\ln P''(Z_n(x))} = 1 \qquad \forall\, x \in X.$$

Example 2.5. Two W-measures which are equal by dimension may be exhibited as follows. Let μ be any non-atomic Markov measure of order 1 on the sequence space $X = A^{\mathbb{N}}$, $2 \leqslant |A| < \infty$. Furthermore let P^1 be any W-measure on A^1 (using the power set as σ-algebra), such that $P^1(\underline{b}) = 0$ for all $\underline{b} \in A^1$ which satisfy $\mu(\underline{b}) = 0$. Then we define, for all $\underline{b} = (b_1, \ldots, b_{n+1}) \in A^{n+1}$ and for all $n \in \mathbb{N}$, a W-measure P on $(X, \underline{X})$ by letting

$$P(\underline{b}) := P^l(b_1, \ldots, b_l) \cdot \Pi_{i=1}^{n} \frac{\mu(b_i,\ldots,b_{i+l})}{\mu(b_i,\ldots,b_{i+l-1})} .$$

In general this measure is not invariant, but it possesses stationary transition probabilities and it is a generalized Markov measure of order l. In this situation the following three statements are equivalent:

(1) $\mu \underset{\dim}{=} P$;

(2) $\mu \overset{*}{\sim} P$ (i.e. μ and P are weakly equivalent; see Def. 2.1);

(3) μ and P have the same null cylinders of order l.

In order to establish this we first note that (2) always implies (3). Conversely, (2) follows from the assertion (3) in the case under consideration since the probability has been defined by means of the transition probabilities of μ. (In general, (3) is weaker than (2)). But if (2) is valid, we have agreed to let $\frac{\mu}{P}(x) = \frac{P}{\mu}(x) = 1$ for all $x \in X$ contained in a μ-null-cylinder. At the remaining points $x \in X$ we consider the representations (compare § 1.B, (W7))

$$\frac{1}{n}\ln P(Z_{n+l}(x)) = \frac{1}{n}\ln P(Z_l(x)) - E^{l+1}(h_n(x), \mu)$$

and

$$\frac{1}{n}\ln \mu(Z_{n+l}(x)) = \frac{1}{n}\ln \mu(Z_l(x)) - E^{l+1}(h_n(x), \mu),$$

where the first equation is true because P was defined by means of the transition probabilities of the measure μ. Here we have $|\ln P(Z_l(x))| < \infty$, $|\ln \mu(Z_l(x))| < \infty$ and $0 < \alpha < E^{l+1}(h_n(x), \mu) < \infty$ for all $n \in \mathbb{N}$, where $\alpha \in \mathbb{R}^+$ is fixed. By forming quotients we verify that here, too, $\frac{\mu}{P}(x) = \frac{P}{\mu}(x) = 1$. Hence assertion (1) follows from (2). If (2) is not valid then there exists a point $x \in X$ such that $\mu(Z_n(x)) > 0$ for all $n \in \mathbb{N}$ and $P(Z_l(x)) = 0$ (keeping in mind that (3) is then not valid either). At this point we have $\frac{\mu}{P}(x) = 0$. Hence the W-measures μ and P can not be equal by dimension, i.e. (1) also implies (2).//

So far we have considered the P-dimension of a fixed non-atomic W-measure P only. However the concepts which we have introduced and the connections which we have established make it also possible to compare P-dimensions relative to different W-measures P:

Theorem 2.15. a) Let $P', P'' \in \Pi_{na}$, then the following assertion holds:

$$P' \underset{\dim}{\leqslant} P'' \Rightarrow P'\text{-dim} \geqslant P''\text{-dim}.$$

b) Let $P' = \Sigma_{i \in N}\alpha_i P_i$ be a denumerable, non-trivial convex combination of non-atomic W-measures P_i on $(X, \underline{X})$. Then the inequality

$$P'\text{-dim} \geqslant \sup_{i \in N} P_i\text{-dim}$$

holds.

Proof. a) From $P' \underset{\dim}{\leqslant} P''$ we first obtain by Theorem 2.14 the relation $\frac{P''}{P'} \geqslant 1$. Together with Theorem 2.9 this implies the assertion.

b) Each P_i is contained in the face of P; hence P is less or equal by dimension to each P_i. Together with part a) which we have already established this implies the assertion.//

For the time being this ends our study of the relation $\underset{\dim}{\leqslant}$. We shall return to it in Sections 4 and 5. In the following example we compute the P-dimension of a specific set relative to different W-measures P. This example shows that even for finite non-trivial convex combinations $\Sigma\alpha_i P_i$, the symbol "$\geqslant$" in Theorem 2.15 can not be replaced by "=" in general.

Example 2.6. Let A := {0,1,2,3} and let X be the sequence space $A^{\mathbb{N}}$. This defines a dimension system $(X, \{Z_n\})$ as explained in Example 2.1. Now let P_1 and P_2 be two Bernoulli measures (compare § 1.B) on $(X, \underline{X})$ satisfying

$$P_1(0) = P_1(1) = P_2(2) = P_2(3) = \frac{1}{3}$$

and

$$P_2(0) = P_2(1) = P_1(2) = P_1(3) = \frac{1}{6}\ .$$

Furthermore let $y = (y_1, y_2, \ldots)$ be a 0-1-sequence satisfying

$$\liminf_{n\to\infty} h_n(y, 1) = \frac{1}{5} \quad \text{and} \quad \limsup_{n\to\infty} h_n(y, 1) = \frac{4}{5}.$$

(Concerning $h_n(y, 1)$ see § 1.B; such a sequence is obtained, e.g. if each block of zeros is followed by a block of ones whose length is four times larger, for example, $\{1\} \times \{0\}^4 \times \{1\}^{16} \times \{0\}^{64} \times \ldots$). Finally let

$$A_0 := \{0,1\}, \quad A_1 := \{2,3\}$$

and

$$B := \{x = (x_1, x_2, \ldots) \in X \mid x_i \in A_{y_i} \quad \forall\ i \in \mathbb{N}\} = \prod_{i=1}^{\infty} A_{y_i}.$$

Now let us compute P_1-dim(B), P_2-dim(B) and $\frac{1}{2}(P_1+P_2)$-dim(B). For this purpose we introduce two W-measures μ_0 and μ_1 on the set A satisfying

$$\mu_0(0) = \mu_0(1) = \mu_1(2) = \mu_1(3) = \frac{1}{2}$$

and

$$\mu_1(0) = \mu_1(1) = \mu_0(2) = \mu_0(3) = 0.$$

Clearly $\mu_i(A_i) = 1$ for i = 1 and i = 2. Hence we have, using the product measure

$\mu := \prod_{i=1}^{\infty} \mu_{y_i}$ on $(X, \underline{X})$, $\mu(B) = 1$ and

$$\frac{1}{n}\ln \mu(Z_n(x)) = -\ln 2 \quad \forall\ x \in B.$$

On the other hand,

$$\frac{1}{n}\ln P_1(Z_n(x)) = h_n(y,0)\cdot\ln\frac{1}{3} + h_n(y,1)\cdot\ln\frac{1}{6} \qquad \forall\, x \in B.$$

By forming quotients we obtain from the last two equations the relation

$$\frac{\mu}{P_1}(x) = \liminf_{n\to\infty} \frac{\ln 2}{h_n(y,0)\ln 3 + h_n(y,1)\ln 6}$$

$$= \frac{\ln 2}{(1/5)\ln 3 + (4/5)\ln 6} = \frac{\ln 2}{\ln 3 + (4/5)\ln 2}$$

for each point $x \in B$. Hence we have by Theorems 2.8 and 2.12,

$$P_1\text{-}\dim(B) = \frac{\ln 2}{\ln 3 + (4/5)\ln 2}.$$

In the same way it is shown that

$$P_2\text{-}\dim(B) = \frac{\ln 2}{\ln 3 + (4/5)\ln 2}.$$

Now let $P := \frac{1}{2}P_1 + \frac{1}{2}P_2$. For brevity we let $k := k(n) = n \cdot h_n(y, 0)$. For each point $x \in B$ and each $n \in \mathbb{N}$ we then obtain

$$P(Z_n(x)) = \frac{1}{2}P_1(Z_n(x)) + \frac{1}{2}P_2(Z_n(x))$$

$$= \frac{1}{2}\left(\frac{1}{3}\right)^k\left(\frac{1}{6}\right)^{n-k} + \frac{1}{2}\left(\frac{1}{6}\right)^k\left(\frac{1}{3}\right)^{n-k}$$

$$= \frac{1}{2}\left(\frac{1}{3}\right)^n \left(\frac{1}{2}\right)^{n-k} + \left(\frac{1}{2}\right)^k$$

$$\geq \frac{1}{2}\left(\frac{1}{3}\right)^n\left(\frac{1}{2}\right)^{n/2}.$$

Consequently,

$$\liminf_{n\to\infty} \frac{1}{n}\ln P(Z_n(x)) \geq \ln\frac{1}{3} + \frac{1}{2}\ln\frac{1}{2}.$$

By considering a sequence $\{n_i\}$ of natural numbers satisfying $\lim_{i\to\infty} h_{n_i}(y, 0) = \frac{1}{2}$ we then see that the lower bound just obtained is indeed the lower limit of the sequence $\{\frac{1}{n}\ln P(Z_n(x))\}$. Hence we have for the W-measure μ introduced above,

$$\frac{\mu}{P}(x) = \frac{\ln 2}{\ln 3 + (1/2)\ln 2} \qquad \forall\, x \in B.$$

Thus we obtain from Theorems 2.8 and 2.12 the relation

$$P\text{-}\dim(B) = \frac{\ln 2}{\ln 3 + (1/2)\ln 2}.$$

Now it can be verified that

$$\frac{1}{2}(P_1+P_2)\text{-}\dim(B) > \max\{P_1\text{-}\dim(B),\ P_2\text{-}\dim(B)\}.$$

The following theorem describes a special case in which equality holds in Theorem 2.15.b). This special case may occur, e.g., for Markov measures P (compare Chapter II, § 6.C).

<u>Theorem 2.16.</u> Let $P = \Sigma_{i\in N}\, \alpha_i P_i$ be a non-trivial convex combination of non-atomic

W-measures P_i on $(X, \underline{X})$ and suppose there exists a decomposition $\{X_i \mid i \in N\}$ of X with the following property:

$$\forall\, i \in N \quad \forall\, x \in X_i \quad \exists\, n \in \mathbb{N} : P(Z_n(x)) = \alpha_i P_i(Z_n(x)).$$

Then

$$P\text{-dim}(M) = \sup_{i \in N} P_i\text{-dim}(M) \qquad \forall\, M \subset X.$$

Proof. Let $M \subset X$. Then it follows from the assumption that

$$\frac{\mu}{P}(x) = \frac{\mu}{P_i}(x) \qquad \forall\, x \in X_i \qquad \forall\, \mu \in \Pi$$

and therefore

$$P\text{-dim}(M \cap X_i) = P_i\text{-dim}(M \cap X_i)$$

for each $i \in N$. On the other hand, each point $x \in X \setminus X_i$ is contained in some P_i-null cylinder and thus

$$P_i\text{-dim}(M \cap X_i) = P_i\text{-dim}(M).$$

Therefore,

$$P\text{-dim}(M) = \sup_{i \in N} P\text{-dim}(M \cap X_i) = \sup_{i \in N} P_i\text{-dim}(M \cap X_i) = \sup_{i \in N} P_i\text{-dim}(M),$$

which establishes the assertion.//

§ 3. Connections between P-dimension, Billingsley dimension and Hausdorff dimension

In this section we always assume a fixed dimension system $(X, \{Z_n\})$ (see Def. 2.1) and a non-atomic W-measure P on $(X, \underline{X})$ to be given. Then in addition to the P-dimension, the Billingsley dimension "$\text{b-dim}_P(\cdot)$" of subsets of X (see § 2, Introduction) is also defined. First we state the following lemma which gives a simplified definition of the Billingsley dimension.

Lemma 3.1. For any subset M of X,

$$\text{b-dim}_P(M) = \inf\{\delta > 0 \mid P^{\delta}(M) = 0\}.$$

Proof. (For the symbol $P^{\delta}(M)$ see Def. 2.5; for the functions $L_P(M, \delta, \varepsilon)$ and $L_P(M, \delta)$ see the introduction of § 2). Let $M \subset X$ and $\delta > 0$. Then the relation $L_P(M, \delta) = 0$ implies $L_P(M, \delta, \varepsilon) = 0$ for all $\varepsilon > 0$; in particular, $L_P(M, \delta, 1) = P^{\delta}(M) = 0$. On the other hand, if $P^{\delta}(M) = 0$, then there exists, for every $\varepsilon > 0$, a covering $\{A_i\}_{i \in N}$ of M by cylinders A_i such that $\Sigma_{i \in N} P(A_i)^{\delta} < \varepsilon^{\delta}$. But then $\{A_i\}_{i \in N}$ is a P-ε-covering of M, and thus $L_P(M, \delta, \varepsilon) < \varepsilon^{\delta}$. Since $\varepsilon > 0$ was arbitrary, this implies that $L_P(M,\delta) = 0$. In summary we have:

$$L_P(M, \delta) = 0 \iff P^{\delta}(M) = 0 \qquad \forall\, \delta > 0,$$

from which the assertion follows immediately.//

The following direct comparison between P-dimension and Billingsley dimension is possible on the basis of Corollary 2.11.2 and Lemma 3.1.

Theorem 3.1. The inequality

$$\text{b-dim}_P \leq \text{P-dim}$$

is always valid.

The following definition imposes conditions on the dimension system $(X, \{Z_n\})$ under which the converse of Theorem 2.11 holds. If this is the case then equality between P-dimension and Billingsley dimension follows by means of the representation of the latter as given by Lemma 3.1.

Definition 3.1. The dimension system $(X, \{Z_n\})$ is called complete if each monotonically decreasing sequence of cylinders has a non-empty intersection. The dimension system $(X, \{Z_n\})$ is called P-complete if each monotonically decreasing sequence of cylinders with positive P-measure has a non-empty intersection.

Remark 3.1.1. If the dimension system $(X, \{Z_n\})$ is complete then it is, in particular, P-complete.

Remark 3.1.2. The sequence space $A^{\mathbb{N}}$ with denumerable state space, equipped with the structure induced by the decompositions of Example 2.1, is a complete dimension system.

Remark 3.1.3. The property "P-complete" is somewhat stronger than Condition (C) of Billingsley [10] which is occasionally stipulated there and which allows countably many monotonically decreasing sequences of cylinders with positive P-measure to have an empty intersection.

Theorem 3.2. Let the dimension system $(X, \{Z_n\})$ be P-complete. Then every subset M of X satisfies the assertion:

$$P^{\delta}(M) = 0 \Rightarrow \text{P-dim}(M) \leq \delta \qquad \forall\, \delta > 0.$$

Proof. 1. We first sharpen the assumptions. Let $M \subset X$ and $\delta > 0$ with $P^{\delta}(M) = 0$ be given. Then we have to show that $\text{P-dim}(M) \leq \delta$. If $\delta \geq 1$ then this is trivial. Furthermore we note that those subsets of M which can be covered by P-null-cylinders do not make any contribution to the P-dimension of M. Therefore we may assume without loss of generality that $\delta < 1$ and

$$P(Z_n(x)) > 0 \qquad \forall\, x \in M \qquad \forall\, n \in \mathbb{N}.$$

Now we have to construct a W-measure μ such that

$$\mu\text{-P-dim}(M) \leq \delta.$$

2. Next we define recursively a sequence $\{\underline{B}_n\}$ of families of cylinders. Let

$$\underline{B}_0 := \{X\}.$$

Now let the family $\underline{B}_n$ of cylinders be defined for some $n \in \mathbb{N}_0$ such that it has the following properties:

(1) The cylinders of $\underline{B}_n$ are pairwise disjoint and at least of order n;

(2) $M \cap B \neq \emptyset$ and $P(B) > 0 \qquad \forall\, B \in \underline{B}_n$;

(3) $M \subset \cup \underline{B}_n$.

In the relation (3) and in the sequel the symbol $\cup H$ denotes the union of all elements of the set H. For each cylinder $B \in \underline{B}_n$ we have $P^{\delta}(M \cap B) = 0$. Hence there exists a denumerable family $\{A_i^B \mid i \in N_B \subset \mathbb{N}\}$ of disjoint cylinders such that

(4) $M \cap B \subset \cup\{A_i^B \mid i \in N_B\}$;

(5) $M \cap B \cap A_i^B \neq \emptyset$ and $P(A_i^B) > 0 \qquad \forall\, i \in N_B$;

(6) $\sum_{i \in N_B} P(A_i^B)^\delta < P(B)^\delta$.

We then define again a family of cylinders with properties (1), (2) and (3) by letting

$$\underline{B}_{n+1} := \{A_i^B \mid i \in N_B,\ B \in \underline{B}_n\}.$$

3. Next we define a set function μ. First let us put

(7) $\mu(X) := 1$.

If μ has been defined on the cylinders of the family $\underline{B}_n$ for some $n \in \mathbb{N}_0$ then let

$$(8)\quad \mu(A_i^B) := \frac{P(A_i^B)^\delta}{\sum_{j \in N_B} P(A_j^B)^\delta} \cdot \mu(B) \qquad \forall\, i \in N_B \qquad \forall\, B \in \underline{B}_n.$$

Using (7) and applying (8) recursively we obtain a definition of the set function μ for all cylinders which occur in the families $\underline{B}_n$. If A is an arbitrary union of cylinders of order n then we let

$$\mu(A) := \sum_{\substack{B \subset A \\ B \in \underline{B}_n}} \mu(B).$$

In view of (1) and (8) these definitions are compatible. Thus we have introduced the set function μ as a W-measure on the σ-algebra $\underline{A}_n$ generated by the cylinders of order n (for each $n \in \mathbb{N}$); hence μ is a normalized non-negative and additive set function on the algebra $\underline{A}$ generated by all σ-algebras $\underline{A}_n$.

4. In order to establish the zero-continuity of μ on $\underline{A}$, let $\varepsilon > 0$ and let $\{C_n\}$ be a monotonically decreasing sequence of $\underline{A}$-measurable subsets of X satisfying

$$\mu(C_n) > \varepsilon \qquad \forall\, n \in \mathbb{N}.$$

Then we also have

$$\Sigma_{D \in Z_1} \mu(D \cap C_n) > \varepsilon \quad \forall\, n \in \mathbb{N}.$$

Since the sequence $\mu(D \cap C_n)$ is monotonically decreasing for each $D \in Z_1$, there exists a cylinder $D_1 \in Z_1$ and an $\varepsilon_1 > 0$ such that

$$\mu(D_1 \cap C_n) > \varepsilon_1 \qquad \forall\, n \in \mathbb{N}.$$

Now we apply the same argument to the monotonically decreasing sequence $\{D_1 \cap C_n\}$, thus obtaining a cylinder $D_2 \in Z_2$ and a number $\varepsilon_2 > 0$, etc.. This process yields a monotonically decreasing sequence of cylinders $D_1, D_2, \ldots, D_m \in Z_m$, and a sequence of numbers $\varepsilon_m > 0$ such that

$$\mu(D_m \cap C_n) > \varepsilon_m > 0 \qquad \forall\, n \in \mathbb{N} \qquad \forall\, m \in \mathbb{N}.$$

In particular, we always have $\mu(D_m) > 0$ and thus $P(D_m) > 0$ by the definition of μ. In this situation the P-completeness of the dimension system $(X, \{Z_n\})$ guarantees that

$$\bigcap_{m \in \mathbb{N}} D_m \neq \emptyset.$$

Therefore there exists a point $x \in \bigcap_{m \in \mathbb{N}} D_m$. For fixed $n \in \mathbb{N}$ we have $C_n \in \underline{A}_m$ and thus $D_m \subset C_n$ for all sufficiently large $m \in \mathbb{N}$. Consequently we also have

$$x \in \bigcap_{n \in \mathbb{N}} C_n \neq \emptyset.$$

This proves the zero continuity of the set function μ on $\underline{A}$. Hence there exists a unique extension of μ to a W-measure on $(X, \underline{X})$ (see, e.g., Bauer [3], Theorems 3.2, 5.2 and 5.5).

5. In order to show that μ-P-dim(M) $\leqslant \delta$, we observe that, trivially,

$$\mu(X) \geqslant P(X)^{\delta}.$$

If the relation $\mu(B) \geqslant P(B)^{\delta}$ is satisfied for some cylinder $B \in \underline{B}_n$, then it is, in view of (6) and (8), also valid for each cylinder of the family $\{A_i^B \mid i \in N_B\}$. Hence we obtain by induction,

$$\mu(B) \geqslant P(B)^{\delta} \qquad \forall\, B \in \underline{B}_n \qquad \forall\, n \in \mathbb{N},$$

which implies

$$(9) \qquad \frac{\ln \mu(B)}{\ln P(B)} \leqslant \delta \qquad \forall\, B \in \underline{B}_n \qquad \forall\, n \in \mathbb{N}.$$

Now let $x \in M$ be any point. Then there exists because of (3) a sequence $\{n_k\}$ of positive integers such that

$$Z_{n_k}(x) \in \underline{B}_k \qquad \forall\, k \in \mathbb{N}.$$

Hence we obtain from (9)

$$\frac{\mu}{P}(x) = \liminf_{n\to\infty} \frac{\ln \mu(Z_n(x))}{\ln P(Z_n(x))} \leqslant \liminf_{k\to\infty} \frac{\ln \mu(Z_{n_k}(x))}{\ln P(Z_{n_k}(x))} \leqslant \delta.$$

Inasmuch as this is true for any $x \in M$, we obtain

$$\mu\text{-P-dim}(M) \leqslant \delta.$$

Therefore it follows from the definition of the P-dimension (Def. 2.4) that

$$\text{P-dim}(M) \leqslant \delta,$$

which completes the proof.//

The following theorem characterizes the Billingsley dimension "b-$\dim_P(\cdot)$" in P-complete dimension systems as infimum of μ-P-dimensions. This result shall enable us in Chapter II to consider Billingsley dimensions as P-dimensions and thus to apply the insights of this chapter, especially those of Sections 2 and 4. This fact may justify the extensive investigation of P-dimensions in which we are engaged in the present chapter.

Theorem 3.3. If $(X, \{Z_n\})$ is a P-complete dimension system then

$b\text{-}\dim_P = P\text{-}\dim$.

Proof. Combining Corollary 2.11.2 and Theorem 3.2 we obtain the same representation for P-dimensions in P-complete dimension systems as we have established in Lemma 3.1 for Billingsley dimensions.//

Remark 3.2.1. The condition of P-completeness is not very important for the usefulness of Theorem 3.3 for, if the dimension system $(X, \{Z_n\})$ is not P-complete then it can always be embedded in a complete dimension system $(X^*, \{Z_n^*\})$. This is accomplished by letting

$$X^* := \{\{A_n\}_{n\in\mathbb{N}} \mid A_n \in Z_n, A_{n+1} \subset A_n \qquad \forall n \in \mathbb{N}\},$$

by assigning to each cylinder $B \in Z_m$ the cylinder

$$B^* := \{\{A_n\} \in X^* \mid A_m = B\},$$

and finally by forming the decompositions

$$Z_n^* := \{B^* \mid B \in Z_n\} .$$

The embedding $\phi: X \to X^*$ which we have mentioned is then obtained as

$$\phi(x) := \{Z_n(x)\} \in X^* \qquad \forall x \in X.$$

This mapping ϕ is $\underline{X}$-$\underline{X}^*$-measurable (where $\underline{X}^*$ is, of course, the σ-algebra generated by the decompositions Z_n^*). For the image $\mu^* = \phi(\mu) \in \Pi^*$ of a W-measure $\mu \in \Pi$ one has

$$\mu^*(B^*) = \mu(B) \qquad \forall B \in \underline{Z};$$

$$\mu^* \in \Pi_{na}^* \iff \mu \in \Pi_{na}.$$

To each covering $\{A_i \mid i \in N\}$ of a subset M of X by cylinders $A_i \in \underline{Z}$ there corresponds a cylinder covering $\{A_i^* \mid i \in N\}$ of the set $\phi(M)$ in such a way that the inverse correspondence is unique. Therefore we have

$$P^\delta(M) = P^{*\delta}(\phi(M)) \qquad \forall \delta > 0.$$

Hence it follows by means of Lemma 3.1 and Theorem 3.3 that

$$b\text{-}\dim_P(M) = b\text{-}\dim_{P*}(\phi(M)) = P^*\text{-}\dim(\phi(M)).$$

This shows that the Billingsley dimension can always be completely described by means of a P-dimension.

Remark 3.2.2. It is possible to prove Theorems 2.2 and 2.4 of Billingsley [10] by completing the dimension system under consideration according to Remark 3.2.1, then using Theorems 2.9 and 2.10, respectively, making the transition to Billingsley dimensions and then returning to the original dimension system.

Remark 3.2.3. The completion described in Remark 3.2.1 makes it plausible that $b\text{-}\dim_P(M)$ of a set M may be smaller than P-dim(M): It is true that, after making the dimension system complete, we have $\mu^*\text{-}P^*\text{-}\dim(\phi(M)) = \mu\text{-}P\text{-}\dim(M)$ for every subset M of X and every W-measure μ on $(X, \underline{X})$, but it is not true that every W-measure on $(X^*, \underline{X}^*)$ should necessarily be the image μ^* of some W-measure $\mu \in \Pi$. The infimum of the ν-P^*-dimensions over all $\nu \in \Pi^*$ may therefore be smaller than the infimum of the μ-P-dimensions, extended over all $\mu \in \Pi$. Nevertheless, the author is not aware of any example in which $b\text{-}\dim_P(M) < P\text{-}\dim(M)$. In other words, he has not been able to determine to what extent the assumption of P-completeness may be weakened in Theorem 3.2 or whether it may perhaps be abandoned entirely.

Remark 3.2.4. The following would be a possibility to enforce equality between P-dimension and Billingsley dimension: Instead of W-measures only, we might admit more general set functions μ, defined for each $n \in \mathbb{N}$ on the σ-algebra $\underline{A}_n$ generated by the decomposition Z_n, provided μ is a W-measure on that σ-algebra. Let Θ be the set of all these set functions. Under the completion procedure described in Remark 3.2.1, each set function $\mu \in \Theta$ determines a set function μ^* which may be extended to a W-measure on $(X^*, \underline{X}^*)$, and each W-measure on $(X^*, \underline{X}^*)$ stems from some $\mu \in \Theta$ in this manner since the set functions $\mu \in \Theta$ have been chosen accordingly. Thus, letting

$$\Theta\text{-}P\text{-}\dim(M) := \inf_{\mu\in\Theta} \mu\text{-}P\text{-}\dim(M) \qquad \forall\, M \subset X,$$

we have

$$b\text{-}\dim_P(M) = P^*\text{-}\dim(\phi(M)) = \Theta\text{-}P\text{-}\dim(M) \quad \forall\, M \subset X.$$

In this fashion we would circumvent the procedure of completing the dimension system, and the Θ-P-dimension would be a concept similar to the P-dimension and equivalent to Billingsley dimension.

We are now in a position to express the Hausdorff dimension of sets of real numbers in terms of their P-dimension, using Satz 2 of Wegmann [51] which establishes a connection between Billingsley and Hausdorff dimensions (h-dim) of such sets. First we state the following lemma which expresses an apparently known representation of the Hausdorff dimension which is analogous to the representation given for Billingsley dimensions by Lemma 3.1.

Lemma 3.2. The Hausdorff dimension of a subset M of the interval [0,1] may be computed as follows: For each $\alpha > 0$ we let

$$\lambda^*(M, \alpha) := \inf \sum_{i\in N} \lambda(A_i)^\alpha,$$

where λ denotes Lebesgue measure on $\mathbb{R}$, and the infimum is extended over all denumerable coverings of M by intervals A_i. Then we have

$$h\text{-}\dim(M) = \inf \{\alpha > 0 \mid \lambda^*(M, \alpha) = 0\}.$$

We omit the proof of Lemma 3.2 since it is analogous to the proof of Lemma 3.1. For some purposes it may be advantageous to use Lemma 3.2 as definition of the Hausdorff dimension of subsets of the interval [0,1]. In our terminology, Wegman's theorem assumes the following form:

Theorem. (Wegmann [51], Satz 2). Let $([0,1], \{Z_n\})$ be a dimension system on the unit interval, let $M \subset [0,1]$ and assume the following conditions to be satisfied:

(A) Each cylinder is an interval;

(B) $\lim_{n\to\infty} \lambda(Z_n(x)) = 0 \qquad \forall\, x \in X;$

(C) $\lim_{n\to\infty} \dfrac{\ln \lambda(Z_n(x))}{\ln \lambda(Z_{n+1}(x))} = 1 \qquad \forall\, x \in M.$

Then $\text{h-dim}(M) = \text{b-dim}_P(M)$.

Wegmann [51] proves this theorem by verifying the assumptions of Satz 3 of his paper [50] which establishes a connection between the Hausdorff dimensions induced by two different metrics on a set. We give a direct proof which does not use additional results.

Proof. 1. Clearly one has

$$\text{h-dim}(M) \leqslant \text{b-dim}_P(M)$$

since the intervals which are admissible for computing the right-hand side by Lemma 3.1 form a sub-family of those available for computing the left-hand side by Lemma 3.2.

2. In order to show that $\text{h-dim}(M) \geqslant \text{b-dim}_P(M)$ we let

$$M' := \{x \in M \mid \lambda(Z_n(x)) > 0 \quad \forall\, n \in \mathbb{N}\} \cap (0,1).$$

Since the set $M \setminus M'$ is denumerable it suffices to show

$$\text{h-dim}(M') \geqslant \text{b-dim}_P(M').$$

For each $\varepsilon \in (0,1)$ and $l \in \mathbb{N}$ let M_l^ε be the set of all points $x \in M'$ satisfying

(1) $\quad \lambda(Z_n(x)) < \frac{1}{l} \Rightarrow \lambda(Z_{n-1}(x)) < \lambda(Z_n(x))^{1-\varepsilon} \qquad \forall\, n \in \mathbb{N}.$

In view of Conditions (B) and (C) and the choice of the subset M' we have

(2) $\quad M' = \cup_{l\in\mathbb{N}} M_l^\varepsilon \qquad \forall\, \varepsilon \in (0,1).$

Now let $\varepsilon \in (0,1)$, $l \in \mathbb{N}$ and $s > \text{h-dim}(M)$ be fixed. Then we also have $s > \text{h-dim}(M_l^\varepsilon)$, and there exists, for any $t \in (0, l^{-s})$, a denumerable covering $\{(a_i, b_i) \mid i \in \mathbb{N}\}$ of M_l^ε by open intervals (a_i, b_i) which are completely contained in $[0,1]$ and which satisfy the inequality

(3) $\quad \Sigma_{i\in N} (b_i - a_i)^s < t.$

Now let $i \in \mathbb{N}$ also be an arbitrary but fixed number. Then there exists, by Condition (B), for each point $x \in M_l^\varepsilon \cap (a_i, b_i)$, a smallest number $n(x) \in \mathbb{N}$ such that

(4) $\quad Z_{n(x)}(x) \subset (a_i, b_i).$

Thus

(5) $\quad \lambda(Z_{n(x)}(x)) \leqslant b_i - a_i < \frac{1}{l},$

and each interval of the form $Z_{n(x)-1}(x)$ contains at least one of the points a_i or b_i. Hence there are at least two maximal intervals B_i^1 and B_i^2 (where possibly $B_i^1 = B_i^2$) of the form $Z_{n(x)-1}(x)$ for which the following two assertions are valid:

(6) $M_1^\varepsilon \cap (a_i, b_i) \subset B_i^1 \cup B_i^2$;

(7) $\lambda(B_i^j) = \lambda(Z_{n(x)-1}(x)) < \lambda(Z_{n(x)}(x))^{1-\varepsilon} < (b_i - a_i)^{1-\varepsilon}$

for $j = 1, 2$ and suitable $x = x(j)$.

It follows from (6) and (7) that $\{B_i^j \mid i \in \mathbb{N}, j = 1, 2\}$ is a covering by cylinders (to be understood as intervals in this case) of the set M_1^ε satisfying

$$\sum_{i \in \mathbb{N}} \lambda(B_i^j)^{s/(1-\varepsilon)} < 2 \cdot \sum_{i \in \mathbb{N}} (B_i - a_i)^s < 2t.$$

Since t may be chosen arbitrarily small, the last inequality implies

$$\text{b-dim}_\lambda(M_1^\varepsilon) \leq \frac{s}{1-\varepsilon}$$

and consequently, by (2), we also have

$$\text{b-dim}_\lambda(M') \leq \frac{s}{1-\varepsilon} .$$

Now letting ε tend to zero and s tend to h-dim(M') from above, we obtain the inequality $\text{b-dim}_\lambda(M') \leq \text{h-dim}(M')$, which finishes the proof.//

Remark 3.3. The conditions of Wegmann's theorem are satisfied, in particular, by the sequence of decompositions which is induced on the interval [0,1] by the g-adic digit representation. Let $g \in \mathbb{N}$, $g \geq 2$, and $A = \{0, 1, 2, \ldots, g-1\}$. Then each number $x \in [0,1]$ has a unique g-adic representation

$$x = \sum_{i=1}^{\infty} \frac{e_i(x)}{g^i} \qquad \text{with } e_i(x) \in A \qquad \forall\, i \in \mathbb{N},$$

where we stipulate that, for all $x \neq 0$, infinitely many digits $e_i(x)$ have to be different from zero. As cylinders of order n we define the sets

$$\{x \in [0,1] \mid e_i(x) = a_i \quad i = 1, 2, \ldots n\} \text{ with } a_i \in A \;\forall\, i = 1, \ldots n.$$

Then the completion of the dimension system $([0,1], \{Z_n\})$ according to Remark 3.2.1 may be identified with the sequence space $A^{\mathbb{N}}$ and the sequence of decompositions introduced in Example 2.1. The embedding

$$\phi : [0,1] \to A^{\mathbb{N}} \quad \text{with} \quad \phi(x) := \{e_i(x)\}_{i \in \mathbb{N}}$$

leads from Lebesgue measure λ to a Bernoulli measure P_g which satisfies

$$P_g(a) = \frac{1}{g} \qquad \forall\, a \in A$$

and is therefore called measure of equidistribution. Now the conditions of Wegmann's theorem are satisfied while, on the other hand, the assumptions of Theorem 3.3 are also valid. Combining both theorems we obtain the following result.

Theorem 3.4. With the notation of Remark 3.3 the following propositions are valid:

a) $h\text{-dim}(M) = P_g\text{-dim}(\phi(M))$ $\quad\forall\, M \subset [0,1]$;

b) $P_g\text{-dim}(M') = h\text{-dim}(\phi^{-1}(M'))$ $\quad\forall\, M' \subset A^{\mathbb{N}}$.

Proof. Proposition b) follows from a) since the set $\phi(\phi^{-1}(M))$ differs from M by at most denumerably many points. Proposition a) is obtained by combining Wegmann's theorem and Theorem 3.3 which yields the chain of equations,

$$h\text{-dim}(M) = b\text{-dim}_\lambda(M) = \lambda^*\text{-dim}(\phi(M)) = P_g\text{-dim}(\phi(M)).//$$

Remark 3.4. For other representations of real numbers (see, e.g., Galambos [15]) analogues to Theorem 3.4 hold provided the conditions of Wegmann's theorem are satisfied. This is the case, for example, if the digit set A is finite and the measure induced on the space $A^{\mathbb{N}}$ is an invariant, non-atomic Markov measure (see also Wegmann [51]).

§ 4. The quasimetric q and the metric q^*

In this section let again $(X, \{Z_n\})$ denote a fixed dimension system (see Definition 2.1). Then $\underline{X}$ is the σ-algebra generated by all cylinders, and Π or Π_{na} denote the families of all W-measures and all non-atomic W-measures on the measurable space $(X, \underline{X})$, respectively.

4.A. The quasimetric q

According to Theorem 2.6.c) there exists, for each set M_α of a family $\{M_\alpha \mid \alpha \in I\}$ of subsets of X, a W-measure μ_α such that

$$P\text{-dim}(M_\alpha) = \mu_\alpha\text{-P-dim}(M_\alpha).$$

If we can now find a W-measure μ with $\mu \underset{\dim}{\leqslant} \mu_\alpha$ for all $\alpha \in I$ (for the symbol " $\underset{\dim}{\leqslant}$ " see Def. 2.6) then we have

$$P\text{-dim}(M_\alpha) = \mu\text{-P-dim}(M_\alpha) \qquad \forall\, \alpha \in I$$

(compare Remark 2.7.7). Hence, by Theorem 2.7,

$$\text{(SUP)} \quad P\text{-dim}(\cup_{\alpha\in I} M_\alpha) = \sup_{\alpha\in I} P\text{-dim}(M_\alpha).$$

This consideration motivates the search for conditions to be imposed on families of W-measures in order to guarantee the existence of a lower bound relative to the partial order $\underset{\dim}{\leqslant}$. (For denumerable families boundedness from below has been established in Remark 2.7.6). This search is facilitated by introducing a metric or quasimetric on Π relative to which the mappings μ-P-dim(M) and P-dim(M) are continuous in μ and P, respectively.

The following definition is motivated by Theorem 2.5:

Definition 4.1. For any two W-measure $\mu, \nu \in \Pi$ a distance $q(\mu, \nu)$ is defined by,

$$q(\mu, \nu) := \sup_{x \in X} \max \{|\ln\frac{\mu}{\nu}(x)|, |\ln\frac{\nu}{\mu}(x)|\}.$$

Lemma 4.1. The distance q is a quasimetric on Π.

Proof. The distance q is always non-negative but it may become infinite. By definition it is symmetric in the two variables. By Theorem 2.1 one always has $1 = \frac{\mu}{\mu} \geqslant \frac{\mu}{\nu} \cdot \frac{\nu}{\mu}$ and thus, taking logarithms,

$$0 \leqslant (-\ln\frac{\mu}{\nu}(x)) + (-\ln\frac{\nu}{\mu}(x)) \qquad \forall\, x \in X.$$

Here the case "$0 \leqslant -\ln 0 + (-\ln\infty)$" has to be considered separately. Hence at most one of the two terms on the right-hand side of the inequality is negative and it does not exceed the other term in absolute value. Hence we also have the equation

$$q(\mu, \nu) = \sup_{x \in X} \max \{-\ln\frac{\mu}{\nu}(x), -\ln\frac{\nu}{\mu}(x)\}.$$

For three W-measures μ, μ', $\mu'' \in \Pi$ we obtain in the same manner

$$-\ln\frac{\mu}{\mu''}(x) \leqslant (-\ln\frac{\mu}{\mu'}(x)) + (-\ln\frac{\mu'}{\mu''}(x)) \leqslant q(\mu, \mu') + q(\mu', \mu''),$$

which yields the triangular inequality

$$q(\mu, \mu'') \leqslant q(\mu, \mu') + q(\mu', \mu'') \qquad \forall\, \mu, \mu', \mu'' \in \Pi$$

by a simple computation. This proves the lemma.//

Remark 4.1.1. An equivalent definition of the distance q could be obtained by letting

$$q(\mu, \nu) = \sup_{x \in X} \limsup_{n \to \infty} |\ln\frac{\ln \mu(Z_n(x))}{\ln \nu(Z_n(x))}| .$$

This identity shows that, given two W-measures with finite q-distance, either both belong to Π_{na} or none of them does. Furthermore, any two W-measure with finite q-distance are weakly equivalent, i.e. they have the same null-cylinders.

Remark 4.1.2. The topology induced by q on Π is, in general, not the weak topology if X is a sequence space $A^{\mathbb{N}}$ (see § 1.B). Topological concepts related to the topology induced by q are therefore labeled as such. In particular, the q-closure of a subset Θ of Π is denoted by $\overline{\Theta}^{q}$.

The following theorem is an immediate consequence of Definition 3.1 and Theorems 2.5 and 2.9:

Theorem 4.1. Let $\mu, \nu \in \Pi$, $P, P' \in \Pi_{na}$ and $\varepsilon > 0$. Then the following assertions are true :

a) $q(\mu, \nu) \leqslant \varepsilon < \infty \Rightarrow e^{-\varepsilon} \cdot \frac{\mu}{P} \leqslant \frac{\nu}{P} \leqslant e^{+\varepsilon} \cdot \frac{\mu}{P}$ or

$q(\mu, \nu) \leqslant \varepsilon < \infty \Rightarrow e^{-\varepsilon} \cdot \mu\text{-}P\text{-dim} \leqslant \nu\text{-}P\text{-dim} \leqslant e^{+\varepsilon} \cdot \mu\text{-}P\text{-dim}$.

b) $q(P, P') \leqslant \varepsilon < \infty \Rightarrow e^{-\varepsilon} \cdot P\text{-dim} \leqslant P'\text{-dim} \leqslant e^{+\varepsilon} \cdot P\text{-dim}$.

Remark 4.2.1. For a fixed set M, the mappings μ-P-dim(M) and P-dim(M) are, by Theorem

4.1, continuous relative to μ and P, respectively, uniformly in M.

Remark 4.2.2. Non-atomic W-measures with finite q-distance assign the dimension 0 to the same sets. If two W-measures have q-distance 0 then they are equal by dimension (see Def. 2.6).

Remark 4.2.3. If two non-atomic W-measures are equal by dimension then they have q-distance 0 by Corollary 2.14.

The usefulness of the distance q for computing dimensions hinges on its compatibility with the partial order $\underset{\dim}{\leq}$ (compare Def. 2.6).

Theorem 4.2. Let $\Theta \subset \Pi$ and $\mu \in \Pi$ be such that

$$\mu \underset{\dim}{\leq} \nu \qquad \forall \nu \in \Theta.$$

Then one has also

$$\mu \underset{\dim}{\leq} \nu \qquad \forall \nu \in \overline{\Theta}^q.$$

Proof. Let $P \in \Pi_{na}$, $\nu \in \overline{\Theta}^q$ and $x \in X$. For each $\varepsilon > 0$ there exists a $\nu' \in \Theta$ such that $q(\nu, \nu') < \varepsilon$. Thus

$$\frac{\mu}{P}(x) \leq \frac{\nu'}{P}(x) \leq e^{\varepsilon} \cdot \frac{\nu}{P}(x),$$

where the first inequality follows from $\mu \underset{\dim}{\leq} \nu'$ and the second one is implied by $q(\nu, \nu') < \varepsilon$ according to Theorem 4.1. Since ε, x and P are arbitrary, it follows that $\mu \underset{\dim}{\leq} \nu$. This completes the proof of the theorem.//

Corollary 4.2. (See Def. 1.1 für the definition of the face $\Sigma(\mu)$ of a W-measure μ). From Remark 2.7.4 combined with Theorem 4.2 we obtain

$$\mu \underset{\dim}{\leq} \nu \qquad \forall \nu \in \overline{\Sigma(\mu)}^q \qquad \forall \mu \in \Pi.$$

Theorem 4.3. Every q-separable family Θ of W-measures $\nu \in \Pi$ is bounded from below relative to the partial order "$\underset{\dim}{\leq}$"; i.e. there exists, for each q-separable set $\Theta \subset \Pi$, a W-measure $\mu \in \Pi$ such that

$$\mu \underset{\dim}{\leq} \nu \qquad \forall \nu \in \Theta.$$

Proof. There exists a sequence $\{\nu_i\}$ of W-measures $\nu_i \in \Theta$ which is q-dense in Θ. Letting $\mu := \Sigma_{i \in \mathbb{N}} 2^{-i} \cdot \mu_i$ we obtain $\mu \underset{\dim}{\leq} \nu_i$ for all $i \in \mathbb{N}$ (compare Corollary 2.2.2), which implies the assertion by Theorem 4.2.//

In the same manner as we have proved Theorem 4.2 one can prove the following proposition by means of Theorem 4.1.b):

Theorem 4.4. Let $\Theta \subset \Pi_{na}$ and $P_o \in \Pi_{na}$ such that

$P_o\text{-dim} \geq P\text{-dim} \qquad \forall P \in \Theta.$

Then one also has

$P_o\text{-dim} \geq P\text{-dim} \qquad \forall P \in \overline{\Theta}^q.$

<u>Corollary 4.4.1.</u> Let $P_o \in \Pi_{na}$. Then

$P_o\text{-dim} \geq P\text{-dim} \qquad \forall P \in \overline{\Sigma(P_o)}^q.$

<u>Corollary 4.4.2.</u> For each q-separable family $\Theta \subset \Pi_{na}$ there exists a W-measure $P_o \in \Pi_{na}$ such that

$P_o\text{-dim} \geq P\text{-dim} \qquad \forall P \in \Theta.$

<u>Example 4.1.</u> Bernoulli measures on $X = A^{\mathbb{N}}$. Let X be the sequence space $A^{\mathbb{N}}$ with denumerable state space A. For any pair μ, ν of Bernoulli measures it follows from § 1.B, (W6), that

$$\inf_{a\in A} \frac{\ln \mu(a)}{\ln \nu(a)} \leq \frac{\ln \mu(Z_n(x))}{\ln \nu(Z_n(x))} \leq \sup_{a\in A} \frac{\ln \mu(a)}{\ln \nu(a)} \qquad \forall x \in X \qquad \forall n \in \mathbb{N}.$$

By means of Remark 4.1.1 this implies that

$$q(\mu, \nu) \leq \sup_{a\in A} \left|\ln \frac{\ln \mu(a)}{\ln \nu(a)}\right| .$$

By substituting points of the form $x = (a, a, \ldots)$ with $a \in A$ we verify that

$$q(\mu, \nu) = \sup_{a\in A} \left|\ln \frac{\ln \mu(a)}{\ln \nu(a)}\right| .$$

If the state space A is finite then the q-distance of two Bernoulli measures is finite if and only if they assign positive probabilities to the same states $a \in A$, i.e. if they are weakly equivalent. Furthermore, if the state space A is finite, then the set of Bernoulli measures with rational $\mu(a)$ for all $a \in A$ is denumerable and q-dense within the set of all Bernoulli measures. Hence, in the case of a finite state space A the Bernoulli measures form a q-separable family.

If the state space A is countably infinite then even the family Θ of all Bernoulli measures μ with

$\mu(a) > 0 \qquad \forall a \in A$

fails to be q-separable. In this case we let $A := \mathbb{N}$ without loss of generality, and we assign to each set R of odd positive integers a Bernoulli measure $\mu_R \in \Theta$ by letting

$$\left.\begin{aligned} \mu_R(n) &:= 2^{-2n} \\ \mu_R(2n) &:= 2^{-n} \end{aligned}\right\} \qquad \forall n \in R$$

$$\mu_R(n) := 2^{-n} \qquad \forall n \in \mathbb{N} \setminus (R \cup 2R)$$

Then we always obtain

$q(\mu_R, \mu_{R'}) = \ln 2$

for any two Bernoulli measures μ_R and $\mu_{R'}$ of this non-denumerable sub-family. Hence the family Θ of Bernoulli measures can not be q-separable.

Example 4.2. Markov measures on $X = A^{\mathbb{N}}$. Let X be the sequence space $A^{\mathbb{N}}$ with denumerable state space A. Each Markov measure μ of order 1 permits, according to § 1.B, (W7), the representation

$$\frac{1}{n}\ln \mu(Z_{n+1}(x)) = \frac{1}{n}\ln \mu(Z_1(x)) + \sum_{\underline{b}\in A^{l+1}} h_n(x\ ,\ \underline{b})\ \ln \mu(\underline{b}/b')$$

for each cylinder $Z_n(x)$, where $0 \leqslant h_n(x,\ b)$ and $\sum_{\underline{b}\in A^{l+1}} h_n(x,\ \underline{b}) = 1$. For the q-distance of two weakly equivalent Markov measures μ and ν this yields the inequality

$$q(\mu,\ \nu) \leqslant \max\ \{ \sup_{\substack{\underline{b}\in A^{l+1}\\ \mu(\underline{b})>0}} \left|\ln \frac{\ln \mu(\underline{b}/\underline{b}')}{\ln \nu(\underline{b}/\underline{b}')}\right|,\ \sup_{\underline{b}\in A^{l}} \left|\ln \frac{\ln \mu(\underline{b})}{\ln \nu(\underline{b})}\right| \}\ ,$$

since the following three cases may occur for the points $x \in X$:

Case 1: x belongs to a μ-null-cylinder. Then x is also contained in a ν-null cylinder and one has $\frac{\mu}{\nu}(x) = \frac{\nu}{\mu}(x) = 1$, the logarithm of which is 0.

Case 2: x does not belong to any μ-null-cylinder and $\lim_{n\to\infty} \mu(Z_n(x)) = 0$. Then both statements are also true for ν, and the relation stated above furnishes the inequality

$$\max\ \{|\ln\frac{\mu}{\nu}(x)|,\ |\ln\frac{\nu}{\mu}(x)|\} \leqslant \sup_{\substack{\underline{b}\in A^{l+1}\\ \mu(\underline{b})>0}} \left|\ln \frac{\ln \mu(\underline{b}/\underline{b}')}{\ln \nu(\underline{b}/\underline{b}')}\right|.$$

Case 3: If $\lim_{n\to\infty} \mu(Z_n(x)) > 0$, then $\mu(Z_n(x)) = \mu(Z_1(x))$ and $\nu(Z_n(x)) = \nu(Z_1(x))$ for all $n \geqslant 1$. Consequently,

$$\max\ \{|\ln\frac{\mu}{\nu}(x)|,\ \ |\ln\frac{\nu}{\mu}(x)|\} \leqslant \sup_{\underline{b}\in A^{l}} \left|\ln \frac{\ln \mu(\underline{b})}{\ln \nu(\underline{b})}\right|.$$

If the state space A is finite then there are, for each $l \in \mathbb{N}_0$, only finitely many classes of weakly equivalent Markov measures of order l in which those denumerable many Markov measures which assign rational probabilities to all cylinders form q-dense sets. This proves the following theorem:

Theorem 4.5. Let X be the sequence space $A^{\mathbb{N}}$ with finite state space A. Then the family of all Markov measures (of arbitrary orders) is q-separable and hence bounded from below relative to the partial ordering " $\underset{\dim}{\leqslant}$ "; i.e. there exists a W-measure $\mu_0 \in \Pi$ such that

$$\mu_0 \underset{\dim}{\leqslant} \mu \qquad \text{for all Markov measures } \mu \in \Pi.$$

This theorem will play an essential role in Sections 7 and 8 in connection with the study of Billingsley dimensions of non-denumerable unions of saturated sets.

4.B. The metric q*

In this part of Section 4 we are going to consider those W-measures μ and P which are expressible as integrals over Markov kernels. This leads to theorems which are similar to those of § 4.A. and to new lower bounds for families of W-measures.

In the sequel let $(Y, \underline{Y})$ be any measurable space and let K be a Markov kernel of $(Y, \underline{Y})$ relative to $(X, \underline{X})$. Thus the following propositions are true (see Bauer [3], § 56):

(K1) K is a mapping of $Y \times \underline{X}$ into $\mathbb{R}_0^+$;

(K2) for fixed $B \in \underline{X}$ the mapping $K(\cdot, B)$ of Y into $\mathbb{R}_0^+$ is measurable;

(K3) for fixed $y \in Y$, $K(y, \cdot) =: \mu_y$ is a W-measure on $(X, \underline{X})$.

To a large extent, Y may be identified with a subset of Π in view of property (K3). In order to facilitate their distinction from W-measures on $(X, \underline{X})$, W-measures on $(Y, \underline{Y})$ shall be called W-distributions in the sequel. If $\tilde{\mu}$ is a W-distribution on $(Y, \underline{Y})$ then a W-measure $\mu := K\tilde{\mu}$ is defined on $(X, \underline{X})$ by letting

$$\mu(B) := \int K(y, B)\, d\tilde{\mu}(y) \qquad \forall\, B \in \underline{X}.$$

The quasimetric q is too weak for establishing connections between the W-measure $\mu = K\tilde{\mu}$ and the W-measure μ_y relative to dimension. Therefore we introduce the following metric q* on Π:

Definition 4.2. For any two W-measures $\mu, \nu \in \Pi$ a distance q* is defined by

$$q^*(\mu, \nu) := \sup_{x \in X} \sup_{n \in \mathbb{N}} \left| \ln \frac{\ln \mu(Z_n(x))}{\ln \nu(Z_n(x))} \right| .$$

Remark 4.3.1. The function q* is a metric on Π. By Remark 4.1.1 we have $q \leq q^*$. Hence the topology induced by q* on Π is finer than the topology induced by q. On certain sub-families of Π, however, the distances q and q* may be equal; this is the case, e.g., on the family of Bernoulli measures on a sequence space with denumerable state space.

Remark 4.3.2. In the definition of q*, the symbol "$\sup_{x \in X}$" may be replaced by "$\sup_{x \in X'}$", where X' is a denumerable subset of X which contains at least one point from each cylinder.

Remark 4.3.3. By defining

$$q^*(y, y') := q^*(\mu_y, \mu_{y'}) \qquad \forall\, y, y' \in Y$$

the metric q* is transferred to Y. In general, the pair (Y, q*) is a quasimetric space only since the mapping $y \to \mu_y$ need not be injective. For each $y \in Y$ and every $\varepsilon > 0$ the q*-neighbourhood

$$U^{q^*}(y, \varepsilon) = \{z \in Y \mid q^*(z, y) < \varepsilon\}$$

of the point y belongs to the σ-algebra $\underline{Y}$ inasmuch as it arises, by Remark 4.3.2, through denumerably many set operations performed on inverse images of intervals under the $\underline{Y}$-measurable mappings $K(\cdot, B)$.

The following definition provides a substitute for the concept of non-trivial convex combinations of W-measures.

Definition 4.3. For a given W-distribution $\tilde{\mu}$ on $(Y, \underline{Y})$, the points $y \in Y$ and the W-measures $\mu_y \in \Pi$ satisfying

$$\tilde{\mu}(U^{q^*}(y, \varepsilon)) > 0 \qquad \forall \varepsilon > 0$$

are called q*-adherent points of $\tilde{\mu}$.

Remark 4.4. If the σ-algebra $\underline{Y}$ contains the σ-algebra of q*-Borel sets, i.e. if each q*-open set is $\underline{Y}$-measurable, then the set of all q*-adherent points of $\tilde{\mu}$ is equal to the spectrum of $\tilde{\mu}$ (see Parthasarathy [35], page 28).

With this terminology the following theorems are true:

Theorem 4.6. Let $\tilde{\mu}$ be a W-distribution on $(Y, \underline{Y})$ and let $\mu := K\tilde{\mu}$. Then each q*-adherent point μ_y of the W-distribution $\tilde{\mu}$ is a q-adherent point of the face $\Sigma(\mu)$, i.e. one has

$$\mu_y \in \overline{\Sigma(\mu)}^q.$$

(See Def. 1.1 for the face $\Sigma(\mu)$).

Proof. Let $\mu_y \in \Pi$ with $y \in Y$ be a q*-adherent point of the W-distribution $\tilde{\mu}$ on $(Y, \underline{Y})$, and let $\varepsilon > 0$. Furthermore, we define

$$Y_1 := U^{q^*}(y, \varepsilon), \ \alpha := \tilde{\mu}(Y_1), \text{ and } \mu_1(B) := \frac{1}{\alpha}\, \tilde{\mu}(B \cap Y_1) \qquad \forall B \in \underline{Y},$$

observing that $\alpha > 0$. Then $\tilde{\mu}_1$ is again a W-distribution on $(Y, \underline{Y})$ and thus

$$\mu_1 := K\tilde{\mu}_1 \in \Pi.$$

In order to show that μ_1 is contained in the face $\Sigma(\mu)$ we distinguish two cases: If $\alpha = 1$, then $\mu_1 = \mu$ and trivially, $\mu_1 \in \Sigma(\mu)$. If $0 < \alpha < 1$ then we let

$$\tilde{\mu}_2(B) := \frac{1}{1-\alpha}\, \tilde{\mu}(B \setminus Y_1) \qquad \forall B \in \underline{Y},$$

thus obtaining a second W-measure $\mu_2 := K\tilde{\mu}_2$ on $(X, \underline{X})$ such that

$$\mu = \alpha\mu_1 + (1 - \alpha)\mu_2.$$

Hence we have $\mu_1 \in \Sigma(\mu)$ also in this case.

For each cylinder $B \in \underline{Z}$ and each point $z \in Y_1$ we have

$$\left|\ln \frac{\ln \mu_y(B)}{\ln \mu_z(B)}\right| < \varepsilon.$$

After solving this inequality for $\mu_z(B)$, integrating with respect to z relative to the W-distribution $\tilde{\mu}_1$ and reconverting the result one obtains

$$\left|\ln \frac{\ln \mu_y(B)}{\ln \mu_1(B)}\right| < \varepsilon \qquad \forall B \in \underline{Z}.$$

Therefore $q^*(\mu_y, \mu_1) < \varepsilon$ and hence $q(\mu_y, \Sigma(\mu)) < \varepsilon$ by Remark 4.3.1. Since $\varepsilon > 0$ is arbitrary, this implies that μ_y is a q-adherent point of the face $\Sigma(\mu)$.//

Theorem 4.7. Let $\tilde{\mu}$ be a W-distribution on (Y, $\underline{Y}$) and $\mu := K\tilde{\mu}$. Then

$$\mu \underset{\dim}{\leqslant} \mu_y$$

for each q*-adherent point μ_y of the W-distribution $\tilde{\mu}$.

Proof. The assertion follows from Theorem 4.6 and Corollary 4.2.//

In the sequel the non-atomic W-measure P is derived from a W-distribution $\tilde{P}$ and the P_y-dimension is compared with the P-dimension.

Theorem 4.8. Let $\tilde{P}$ be a W-distribution on (Y, $\underline{Y}$) such that $\tilde{P}$-almost all W-measures $P_y := K(y,\cdot)$ are non-atomic. Then the measure $P := K\tilde{P}$ is also non-atomic, and every q*-adherent point P_y of the W-distribution $\tilde{P}$ satisfies

$$P\text{-dim} \geqslant P_y\text{-dim}.$$

Proof. For each point $x \in X$ we have by assumption

$$\lim_{n\to\infty} K(y, Z_n(x)) = 0 \qquad \forall\ [\tilde{P}]\ y \in Y.$$

Since convergence is monotonic in this case, integration on y with respect to the W-distribution $\tilde{P}$ shows that the relation

$$\lim_{n\to\infty} P(Z_n(x)) = 0$$

is also valid. Hence the W-measure P is non-atomic. The remaining part of the assertion of the theorem follows from Theorem 4.6 and Corollary 4.4.1.//

In the final part of this section we shall now apply Theorem 4.7 in order to exhibit an example where a lower bound relative to the partial ordering " $\underset{\dim}{\leqslant}$ " is constructed for the family of non-atomic Bernoulli measures over a sequence space with finite state space.

Example 4.3. A lower bound for Bernoulli measures. Let X be the sequence space $A^{\mathbb{N}}$ with the finite state space $A = \{a_1, a_2, \ldots, a_k\}$, $k \geqslant 2$. This determines the dimension system (X, $\{Z_n\}$) and the σ-algebra $\underline{X}$ according to Example 2.1. Furthermore let

$$Y := \{(y_1, y_2, \ldots, y_{k-1}) \in \mathbb{R}^{k-1} \mid y_i \geqslant 0\ \forall\ i = 1, \ldots, k-1,\ \ \Sigma_{i=1}^{k-1} y_i \leqslant 1\};$$

let $\underline{Y}$ be the σ-algebra of the Borel-measurable subsets of Y, and let λ^{k-1} be the (k-1)-dimensional Lebesgue measure on $\mathbb{R}^{k-1}$.

We now assign to each point $y = (y_1, \ldots, y_{k-1}) \in Y$ the Bernoulli measure μ_y which satisfies

$$\mu_y(a_i) = y_i \qquad \forall\ i = 1, \ldots, k-1,$$

$$\mu_y(a_k) = 1 - \Sigma_{i=1}^{k-1} y_i\ .$$

Then the mapping

$$K: Y \times \underline{X} \to \mathbb{R}_0^+ \text{ with } K(y, B) := \mu_y(B) \qquad \forall\ (y, B) \in Y \times \underline{X}$$

is a Markov kernel of $(Y, \underline{Y})$ with respect to $(X, \underline{X})$. Since

$$\lambda^{k-1}(Y) = \frac{1}{(k-1)!},$$

the normalized restriction

$$\tilde{\nu}_k := (k-1)!\ \lambda^{k-1}|_Y$$

is a W-distribution on $(Y, \underline{Y})$ which is transformed into the W-measure

$$\nu_k := K\tilde{\nu}_k$$

on the measurable space $(X, \underline{X})$.

Now let $n \in \mathbb{N}$, let $B \in \underline{Z}$ be a cylinder of order n and let $x \in B$. Then one has $B = Z_n(x)$ and therefore, letting $n_i := n \cdot h_n(x, a_i)$ for all $i = 1, \ldots, k$,

$$\mu_y(B) = \Pi_{i=1}^{k} \mu_y(a_i)^{n_i}$$

(see § 1.B, (W6)) and furthermore

$$\nu_k(B) = (k-1)! \int_Y \mu_y(B)\, d\lambda^{k-1}(y)$$

$$= (k-1)! \int_{y_1=0}^{1} \int_{y_2=0}^{1-y_1} \cdots \int_{y_{k-1}=0}^{1-y_1-\cdots-y_{k-2}} (1-\Sigma_{i=1}^{k-1} y_i)^{n_k}\ \Pi_{i=1}^{k-1} y_i^{n_i}\ dy_{k-1}\cdots dy_1 .$$

Carrying out the integration we obtain

$$\nu_k(B) = \frac{n_1!\ n_2! \ldots n_k!}{n!\ \binom{n+k-1}{k-1}} .$$

As we had to expect, $\nu_k(B)$ does not depend on the order of the states a_i within the cylinder $B = [b_1, b_2, \ldots, b_n]$ but only on their frequencies n_i. Furthermore, the ν_k-measure of the union $C(n_1, n_2, \ldots, n_k)$ of all cylinders B in which the states a_i occur with the same frequencies n_i satisfies

$$\nu_k(C(n_1,\ldots,n_k)) = \frac{1}{\binom{n+k-1}{k-1}} \qquad \text{with } n = \Sigma_{i=1}^{k} n_i .$$

Hence this probability no longer depends on the frequencies n_i. Conversely, ν_k is uniquely determined by the two stated properties. For a Bayes kind of interpretation of the the W-distribution $\tilde{\nu}_k$ and of the W-measure ν_k in statistics see Good [28], especially Chapter 4.

The q*-adherent points of the W-distribution $\tilde{\nu}_k$ are exactly the Bernoulli measures μ_y with $\mu_y(a_i) > 0$ for all $i = 1, \ldots, k$, i.e. exactly those Bernoulli measures which correspond to interior points of Y if considered as a subset of $\mathbb{R}^{k-1}$ under the Euclidean topology. Hence by Theorem 4.7,

$$\nu_k \underset{\dim}{\leq} \mu_y$$

for each Bernoulli measure μ_y corresponding to an interior point y of Y.

By means of Stirling's formula we can show that

$$\lim_{n\to\infty} [\frac{1}{n} \ln \nu_k(Z_n(x)) - \Sigma_{i=1}^{k} h_n(x, a_i) \ln h_n(x, a_i)] = 0 \quad \forall\, x \in X.$$

Since on the other hand

$$\frac{1}{n} \ln \mu_y(Z_n(x)) = \Sigma_{i=1}^{k} h_n(x, a_i) \ln \mu_y(a_i) \leqslant \Sigma_{i=1}^{k} h_n(x, a_i) \ln h_n(x, a_i)$$

for each $y \in Y$, it follows that

$$\frac{\mu_y}{\nu_k}(x) \geqslant 1 \qquad \forall\, x \in X,$$

provided that μ_y is a non-atomic Bernoulli measure. In view of Theorem 2.14 this means that ν_k is a lower bound for all non-atomic Bernoulli measures on the sequence space $X = A^{\mathbb{N}}$ with respect to the partial ordering " $\underset{\text{dim}}{\leqslant}$ ".

§ 5. The P-dimension of probability measures

Let μ be a W-measure on the unit interval [0,1]. Then the minimum of all Hausdorff dimensions of measurable sets M with $\mu(M) = 1$ is studied in the paper Kinney-Pitcher [31] and is called the dimension of the support of μ. In an analogous manner we shall define and investigate in this section the P-dimension of W-measures over a dimension system. At the end we shall also elaborate on the P-dimension of invariant W-measures over the sequence space $A^{\mathbb{N}}$.

The results of this section will not be applied in Chapter II.

Let again $(X, \{Z_n\})$ be a fixed dimension system (see Def. 2.1). Then $\underline{X}$ is the σ-algebra generated by all cylinders, and Π and Π_{na}, respectively, denote the families of W-measures and of non-atomic W-measures on $(X, \underline{X})$. Furthermore let P be an arbitrary, but fixed non-atomic W-measure on $(X, \underline{X})$.

Definition 5.1. For each W-measure $\mu \in \Pi$ let the P-dimension of the W-measure μ be defined by

$$\text{P-dim}(\mu) := \inf \{\text{P-dim}(M) \mid M \in \underline{X},\ \mu(M) = 1\}.$$

Every measurable set $M \subset X$ with $\mu(M) = 1$ and P-dim(M) = P-dim(μ) is called a P-support of μ.

Remark 5.1. One always has $0 \leqslant \text{P-dim}(\mu) \leqslant 1$, where P-dim(P) = 1 by Corollary 2.11.1.

Theorem 5.1. For all $\mu \in \Pi$ the relation

$$\text{P-dim}(\mu) = \mu\text{-ess.sup}_{x \in X} \frac{\mu}{P}(x)$$

is valid.

Proof. Letting $\alpha := \mu\text{-ess.sup}_{x \in X} \frac{\mu}{P}(x)$ and $M_\alpha := \{x \in X \mid \frac{\mu}{P}(x) \leq \alpha\}$, clearly $M_\alpha \in \underline{X}$ and $\mu(M_\alpha) = 1$. Since the μ-essential supremum does not depend on μ-null-sets, it follows from Theorems 2.12 and 2.8 that every measurable set $M \subset X$ with $\mu(M) = 1$ satisfies

$$\text{P-dim}(M) \geq \alpha \geq \mu\text{-P-dim}(M_\alpha) \geq \text{P-dim}(M_\alpha) \geq \alpha.$$

Consequently, $\text{P-dim}(\mu) = \alpha$.//

Remark 5.2.1. In the proof just given the set M_α is a P-support of the W-measure μ.

Remark 5.2.2. It follows from Theorem 5.1 that the value of $\text{P-dim}(\mu)$ is not affected if the space X is enlarged by μ-null-sets. This is the case, e.g., if the dimension system $(X, \{Z_n\})$ is embedded in its completion $(X^*, \{Z_n^*\})$ by the mapping ϕ (notation as in Remark 3.2.1), provided the set $\phi(X)$ is measurable then. Therefore we have in this situation, letting $\mu^* := \phi(\mu)$, $P^* := \phi(P)$ and

$$\text{b-dim}_P(\mu) := \inf \{\text{b-dim}_P(M) \mid M \in \underline{X},\ \mu(M) = 1\} ,$$

the equation

$$\text{b-dim}_P(\mu) = P^*\text{-dim}(\mu^*) = \text{P-dim}(\mu).$$

The following theorem shows that $\text{P-dim}(\mu)$ is large if only the distance between the faces $\Sigma(P)$ and $\Sigma(\mu)$ of the W-measures P and μ (see Def. 1.1) in the sense of the quasi-metric q is small.

Theorem 5.2. For all $\mu \in \Pi$ and $\varepsilon > 0$ the following assertion is valid:

$$q(\Sigma(P), \Sigma(\mu)) \leq \varepsilon \Rightarrow \text{P-dim}(\mu) \geq e^{-\varepsilon}.$$

Proof. According to the assumption there are, for each $\alpha > \varepsilon$, measures $P' \in \Sigma(P)$ and $\mu' \in \Sigma(\mu)$ such that $q(P', \mu') < \alpha$. For each set $M \in \underline{X}$ with $\mu(M) = 1$ we have

$$\text{P-dim}(M) \geq P'\text{-dim}(M) \geq e^{-\alpha} \cdot \mu'\text{-dim}(M) \geq e^{-\alpha} \cdot 1.$$

The first of these inequalities holds because P' is contained in the face $\Sigma(P)$ (compare Corollary 4.4.1), whereas the second one is valid since $q(P', \mu') < \alpha$ (compare Theorem 4.1.b), while the third one is true since the equation $\mu(M) = 1$ implies $\mu'(M) = 1$ because μ' is contained in the face of μ (compare Remark 1.1, (2)). It only remains to add in retrospect that the P'-dimension and the μ'-dimension are well-defined since P' is non-atomic, being contained in the face of P, and μ' is non-atomic according to Remark 4.1.1 because it has a finite q-distance from P'. This chain of inequalities implies the assertion since the number α is arbitrary, subject only to the condition $\alpha > \varepsilon$.//

Corollary 5.2.1. For all $\mu \in \Pi$ and $\varepsilon > 0$ the following is true:

$$q(P, \mu) \leq \varepsilon \Rightarrow \text{P-dim}(\mu) \geq e^{-\varepsilon}.$$

Corollary 5.2.2. For all $\mu \in \Pi$ the following propositions hold:

$$q(\Sigma(P), \Sigma(\mu)) = 0 \Rightarrow \text{P-dim}(\mu) = 1;$$

$$\overline{\Sigma(P)}^q \cap \overline{\Sigma(P)}^q \neq \emptyset \Rightarrow \text{P-dim}(\mu) = 1.$$

Remark 5.3. If $q(P, \mu) \leq \varepsilon$, we even obtain $\mu\text{-ess.inf}_{x \in X} \frac{\mu}{P}(x) \geq e^{-\varepsilon}$, i.e.

$P\text{-dim}(M) \geq e^{-\varepsilon}$ for all $M \in \underline{X}$ with $\mu(M) > 0$.

In addition to Corollary 5.2.2 we may show the following result for the partial ordering "$\underset{\dim}{\leq}$":

Theorem 5.3. For all $\mu \in \Pi$,

$$P \underset{\dim}{\leq} \mu \Rightarrow P\text{-dim}(\mu) = 1.$$

Proof. First we note that the condition $P \underset{\dim}{\leq} \mu$ implies $\mu \in \Pi_{na}$ and furthermore, by Theorem 2.15,

$P\text{-dim}(M) \geq \mu\text{-dim}(M) = 1$

for all measurable subsets M of X satisfying $\mu(M) = 1$. Therefore $P\text{-dim}(\mu) = 1$.//

The following theorem deals with a situation in which the W-measures μ and P are given in terms of integral representations.

Theorem 5.4. For $i = 1, 2$ let $(Y_i, \underline{Y}_i)$ be measurable spaces and let K_i be Markov kernels of $(Y_i, \underline{Y}_i)$ relative to $(X, \underline{X})$. Furthermore let $\tilde{\mu}$ and $\tilde{P}$ be W-distributions on $(Y_1, \underline{Y}_1)$ and $(Y_2, \underline{Y}_2)$ respectively, and let $\mu := K_1\tilde{\mu}$ and $P := K_2\tilde{P}$ (see § 4.B). Suppose there exists a W-measure $\nu \in \Pi$ which is a q*-adherent point of the W-distribution $\tilde{P}$ as well as a q*-adherent point of the W-distribution $\tilde{\mu}$. Then $P\text{-dim}(\mu) = 1$.

Proof. The W-measure ν, being a common q*-adherent point of the W-distributions $\tilde{P}$ and $\tilde{\mu}$, is a common q-adherent point of the faces $\Sigma(P)$ and $\Sigma(\mu)$ (Theorem 4.6). Hence the assumptions of Corollary 5.2.2 are satisfied, which proves the assertion.//

Example 5.1. Let X be the sequence space $A^{\mathbb{N}}$ with the state space $A = \{a_1,...,a_k\}$, $2 \leq k < \infty$. As in Example 4.3 let ν_k denote the invariant measure constructed by means of a W-distribution $\tilde{\nu}_k$ over the Bernoulli measures.

a) As we explained in Example 4.3, a Bernoulli measure μ with

$\mu(a) > 0 \qquad \forall\, a \in A$

is a q*-adherent point of $\tilde{\nu}_k$ and hence by Theorem 4.6 a q-adherent point of the face $\Sigma(\nu_k)$. Therefore Corollary 5.2.2 implies that these Bernoulli measures satisfy

$\nu_k\text{-dim}(\mu) = 1$ and $\mu\text{-dim}(\nu_k) = 1$.

b) For a non-atomic Bernoulli measure μ with

$\mu(a) = 0$ for at least one $a \in A$

ν_k-almost all points $x \in X$ are contained in μ-null cylinders. Therefore we now have

$\mu\text{-dim}(\nu_k) = 0$.

On the other hand, the relation $\nu_k \underset{\dim}{\leq} \mu$ (see Example 4.3) implies by Theorem 5.3 that

ν_k-dim$(\mu) = 1.$

Example 5.2. Ergodic measures on $A^{\mathbb{N}}$. Let X be the sequence space $A^{\mathbb{N}}$ with a finite state space A, let μ be an ergodic W-measure on $A^{\mathbb{N}}$ and P be a non-atomic Markov measure on $A^{\mathbb{N}}$ (see § 1.B). As we have shown in Example 2.4, the equation

$$\text{P-dim}(\mu) = \mu\text{-ess.sup}_{x \in X} \frac{\mu}{P}(x) = \frac{E(\mu)}{E(\mu,P)}$$

holds in this case (see § 1.B for the functions $E(\mu)$ and $E(\mu, P)$).

The following Theorems 5.5 and 5.6 are extensions of Example 5.2. Now let X be the sequence space $A^{\mathbb{N}}$, $2 \leq |A| < \infty$, and let μ be an invariant W-measure on $(X, \underline{X})$. This invariant measure μ induces on the set Π_{inv} of all invariant W-measures a W-distribution $\tilde{\mu}$ (see § 1.B) relative to which almost all W-measures are ergodic. Now we first express the P-dimension of the invariant measure μ in terms of the P-dimension of the ergodic W-measures. This procedure extends the well-known representations for invariant W-measures stated in Lemma 1.2.

Theorem 5.5. Suppose one of the two following conditions is satisfied:

(1) $E(\nu) > 0 \qquad \forall\ [\tilde{\mu}]\ \nu \in \Pi_{inv};$

(2) $\liminf_{n\to\infty} \frac{-n}{\ln P(Z_n(x))} < \infty \qquad \forall\ [\mu]\ x \in X.$

Then $\text{P-dim}(\mu) = \tilde{\mu}\text{-ess.sup}_{\nu \in \Pi_{inv}} [\text{P-dim}(\nu)]$.

Proof. For $\tilde{\mu}$-almost all $\nu \in \Pi_{inv}$ we have by Lemma 1.2

$$\lim_{n\to\infty} \frac{1}{n}\ln \mu(Z_n(x)) = \lim_{n\to\infty} \frac{1}{n}\ln \nu(Z_n(x)) = -E(\nu) \qquad \forall\ [\nu]\ \ x \in X,$$

which implies by means of either Condition (1) or (2) that

$$\frac{\mu}{P}(x) = \frac{\nu}{P}(x) \qquad \forall\ [\nu]\ x \in X.$$

Therefore we obtain by Lemma 1.2 b) the equation

$$\mu\text{-ess.sup}_{x \in X} \frac{\mu}{P}(x) = \tilde{\mu}\text{-ess.sup}_{\nu \in \Pi_{inv}} [\nu\text{-ess.sup}_{x \in X} \frac{\mu}{P}(x)]$$

$$= \tilde{\mu}\text{-ess.sup}_{\nu \in \Pi_{inv}} [\nu\text{-ess.sup}_{x \in X} \frac{\nu}{P}(x)] \quad .$$

In view of Theorem 5.1 this is equivalent to the assertion.//

The W-distribution $\tilde{\mu}$ of the invariant W-measure μ is non-atomic if and only if there exists, for every ergodic measure ν, a measurable set $X_\nu \subset X$ for which $\nu(X_\nu) = 1$ and $\mu(X_\nu) = 0$.

The following theorem states in substance that, on the other hand, μ-dim(X_ν) is large in general.

Theorem 5.6. If the invariant W-measure μ is non-atomic then

$$\mu\text{-dim}(\nu) = 1 \qquad \forall\ [\tilde{\mu}]\ \nu \in \Pi_{inv} \ \text{ with } \ E(\nu) > 0.$$

Proof. For μ-almost all $\nu \in \Pi_{inv}$ with $E(\nu) > 0$ one can show as in the proof of Theorem 5.5 that

$$1 = \frac{\mu}{\mu}(x) = \frac{\nu}{\mu}(x) \qquad \forall\ [\nu]\ x \in X.$$

By Theorem 5.1 this again proves the assertion.//

Remark 5.4. Condition (2) of Theorem 5.5 is satisfied, e.g., if P is a non-atomic Markov measure on the sequence space $A^{\mathbb{N}}$.

CHAPTER II

The Billingsley dimension of saturated sets

Throughout this chapter we denote by A an arbitrary, but fixed finite set with at least two elements; by X the sequence space $A^{\mathbb{N}}$; by P a non-atomic, ergodic Markov measure on the sequence space X. Then the complete dimension system $(X, \{Z_n\})$, the σ-algebra $\underline{X}$ and the set Π of W-measures on $(X, \underline{X})$ and its subsets Π_{na} (non-atomic W-measures), Π_{inv} (W-measures invariant under the shift T) and Π_{erg} (W-measures which are ergodic with respect to the shift T) are well-defined according to § 1.B, Definition 2.1 and Example 2.1.

The main object of this chapter is to determine the Billingsley dimension (now always with respect to the measure P) and hence, by specializing P, also the Hausdorff dimension of the saturated sets mentioned in the Introduction (to be defined exactly in Def. 6.3). This investigation will also furnish a proof of the relation (SUP) for saturated sets which we have announced there.

Since the dimension system $(X, \{Z_n\})$ is complete, P-dimension and Billingsley dimension coincide there by Theorem 3.3. Therefore the Billingsley dimension "$\text{b-dim}_P(M)$" of a subset M of X may always be written as "P-dim(M)", which we shall do in order to preserve congruity with Chapter I. In particular, the results of Sections 2 and 4 may be applied in order to estimate and compute P-dimensions. It shall be discussed in § 6.D to what extent the special W-measure P used in this chapter may be replaced by a more general one.

§ 6. Introductory arguments

6.A. The saturated subsets of X. In this chapter we shall only consider the weak topology (see § 1.B) on the set Π of all W-measures over $(X, \underline{X})$. This topology is induced by the following metric d.

Definition 6.1. For any two W-measures $\mu, \nu \in \Pi$ and any $l \in \mathbb{N}$ let

$$d^l(\mu, \nu) := \max \{|\mu(\underline{b}) - \nu(\underline{b})| \;|\underline{b} \in A^l\}.$$

Then

$$d(\mu, \nu) := \sum_{l \in \mathbb{N}} 2^{-l} \cdot d^l(\mu, \nu).$$

Remark 6.1. By means of this metric and the equations (W1), (W2) and (W3) of § 1.B it is easy to show that (Π, d) is a compact metric space and that Π_{inv} is a closed (and hence compact) convex subset of it. For each $l \in \mathbb{N}$ the function d^l is a quasi-metric on Π.

Given any point $x \in X$, on account of the last remark the sequence $\{h_n(x)\}$ of relative frequencies, where

$$h_n(x)(B) := \frac{1}{n} \cdot \Sigma_{i=0}^{n-1} \chi_B T^i x \qquad \forall\, B \in \underline{X},$$

always possesses limit points in Π.

Definition 6.2. For any point $x \in X$, each limit point $h \in \Pi$ of the sequence $\{h_n(x)\}$ in (Π, d) is called a distribution measure of x. The set of all distribution measures of a point $x \in X$ is denoted by $H(x)$.

In the weak topology the behaviour of the sequence $\{h_n(x)\}$ is determined by the sequences $\{h_n(x, \underline{b})\}$ of all cylinders $\underline{b} \in \underline{Z}$. For a block $\underline{b} \in A^l$ we have

$$h_n(x, \underline{b}) = \frac{1}{n} \cdot |\{i \in \mathbb{N}_0 \mid 0 \leq i \leq n-1,\ (x_{i+1}, x_{i+2}, \ldots, x_{i+l}) = \underline{b}\}|,$$

which is the relative frequency by which the block $\underline{b} = (b_1, b_2, \ldots, b_l)$ occurs within the section $(x_1, x_2, \ldots, x_{n+l-1})$ of the sequence $x = (x_1, x_2, \ldots)$. Hence the sequence $\{h_n(x)\}$ reflects the simultaneous behaviour of the relative frequencies of all blocks. Clearly

$$|h_n(x, \underline{b}) - h_{n+1}(x, b)| \leq \frac{2}{n+1} \qquad \forall\, \underline{b} \in A^l \qquad \forall\, l \in \mathbb{N}$$

and consequently

$$d(h_n(x), h_{n+1}(x)) \leq \frac{2}{n+1} .$$

This implies that the set $H(x)$, being obviously closed, is connected. Now let $h \in H(x)$ be a distribution measure of the point $x \in X$, then there exists, for any $\varepsilon > 0$, a natural number $n > \frac{1}{\varepsilon}$ satisfying $d(h, h_n(x)) < \frac{\varepsilon}{|A|}$. Then, for each block $\underline{b} = (b_1, b_2, \ldots, b_l) \in A^l$,

$$\Big| \sum_{b \in A} h(b, b_1, \ldots, b_l) \;-\; \sum_{b \in A} h(b_1, \ldots, b_l, b) \Big|$$

$$< \Big| \sum_{b \in A} h_n(x, (b, b_1, \ldots, b_l)) - \sum_{b \in A} h_n(x, b_1, \ldots, b_l, b)) \Big| \;+ 2\varepsilon$$

$$= \frac{1}{n} \cdot \Big| \sum_{i=1}^{n} \chi_{[\underline{b}]} T^i x \;-\; \sum_{i=0}^{n-1} \chi_{[\underline{b}]} T^i x \;\Big| + 2\varepsilon < 4\varepsilon .$$

Since $\varepsilon > 0$ may be chosen arbitrarily small, it follows that, since now

$$h(T^{-1}([\underline{b}])) = h(\underline{b}),$$

only invariant W-measures may occur as distribution measures. In summary this result may be stated as the following lemma (compare Volkmann [49] and Colebrook [20]):

Lemma 6.1. For each point $x \in X$ the set $H(x)$ of its distribution measures is a non-empty, closed and connected subset of Π_{inv}.

Now we are in a position to define saturated subsets of X.

Definition 6.3. Let $\mathbb{H}$ denote the set of non-empty, closed, connected subsets of Π_{inv}.

For each $H \in \mathbb{H}$ let

$$M_H := \{x \in X \mid H(x) = H\}.$$

For each subset $\underline{H} \subset \mathbb{H}$ let

$$M_{\underline{H}} := \{x \in X \mid H(x) \in \underline{H}\}.$$

Sets of the form $M_{\underline{H}}$ with $\underline{H} \subset \mathbb{H}$ are called saturated. The sets of the form M_H with $H \in \mathbb{H}$ are called the smallest saturated subsets of X.

For arbitrary H, H' $\in \mathbb{H}$ let

$$\delta(H, H') := \inf \{\varepsilon > 0 \mid H \subset U^d(H', \varepsilon), H' \subset U^d(H, \varepsilon)\}.$$

The power set of $\mathbb{H}$ is denoted by $\underline{\mathbb{H}}$. For any pair $\underline{H}$, $\underline{H}' \in \underline{\mathbb{H}}$ with $\underline{H} \neq \emptyset \neq \underline{H}'$ let

$$\underline{\delta}(\underline{H}, \underline{H}') := \inf \{\varepsilon > 0 \mid \underline{H} \subset U^{\delta}(\underline{H}', \varepsilon), \underline{H}' \subset U^{\delta}(\underline{H}, \varepsilon)\};$$

$$\underline{\delta}(\underline{H}, \emptyset) := \underline{\delta}(\emptyset, \underline{H}) := \infty;$$

$$\underline{\delta}(\emptyset, \emptyset) := 0.$$

Remark 6.2.1. The function δ is the restriction of the Hausdorff metric on the set of compact subsets of Π onto the set of compact, connected subsets of Π_{inv}. Hence δ is a metric on $\mathbb{H}$. The restriction of $\underline{\delta}$ to the set of δ-closed subsets of $\mathbb{H}$ is again a Hausdorff metric; it only is a quasimetric on $\underline{\mathbb{H}}$.

Remark 6.2.2. The sets M_H are the equivalence classes of the equivalence relation

$$x \sim y \quad :\Longleftrightarrow \quad H(x) = H(y) \qquad \forall\, x, y \in X$$

on X. The sets $M_{\underline{H}}$ may be represented in the form

$$M_{\underline{H}} = \bigcup_{H \in \underline{H}} M_H;$$

hence they are unions of equivalence classes and thus they are saturated with respect to the equivalence relation $\sim$.

The following lemma shall be needed for a continuity argument in connection with Theorem 8.4.

Lemma 6.2. ($\mathbb{H}$, δ) is a compact metric space.

Proof. The assertion of the lemma is contained in Theorems 1, § 42.1 and 14, § 46, III of Kuratowski [33].//

6.B. (Π, d) as sequence space. The quasimetric d^1 on Π of Definition 6.1 may also be interpreted as a metric on the set Π^1 of all W-measures on (X, $\underline{A}^1$) or (A^1, $\underline{A}^1$), respectively. Here $\underline{A}^1$ stands for the σ-algebra generated by all cylinders of order 1 over X; upon identifying the cylinders of order 1 with the elements of A^1, the set $\underline{A}^1$ becomes the power set of A^1. Furthermore, Π^1 may also be interpreted as the (k^1-1)-dimensional simplex of all mappings

$$\mu : A^1 \to \mathbb{R}_0^+ \qquad \text{with} \quad \sum_{\underline{b} \in A^1} \mu(\underline{b}) = 1,$$

since a W-measure on (A^1, $\underline{A}^1$) is uniquely determined by its values on the elements of A^l. Now the metric d^l, while not being identical with the Euclidean metric, is nevertheless equivalent to it.

The restriction

$$p^l(\mu) := \mu|_{\underline{A}^l}$$

defines a projection p^l:

$$p^l : \Pi \to \Pi^l \qquad \forall\, l \in \mathbb{N}.$$

In the same manner we define projections $p^l : \Pi^{l'} \to \Pi^l$ for all $l' \geq l$, where the range of the mapping has to be understood from the context. The projections p^l are continuous and affine mappings, i.e., one has

$$p^l(\alpha\mu + (1-\alpha)\nu) = \alpha p^l(\mu) + (1-\alpha)p^l(\nu) \quad \forall\, \mu, \nu \in \Pi \qquad \forall\, \alpha \in [0,1].$$

With each W-measure $\mu \in \Pi$ we can now associate the sequence $\{\mu^l\}$ of W-measures $\mu^l := p^l(\mu)$. This sequence then satisfies the relation

$$\text{(V1)} \quad p^l(\mu^{l+1}) = \mu^l \qquad \forall\, l \in \mathbb{N}.$$

On the other hand, if a sequence $\{\mu^l\}$ of W-measures $\mu^l \in \Pi^l$ is given such that it satisfies the compatibility conditions (V1) then it follows from extension and uniqueness theorems of measure theory that there exists exactly one W-measure $\mu \in \Pi$ with

$$\mu^l = p^l(\mu) \qquad \forall\, l \in \mathbb{N}.$$

We shall then use the notation

$$\mu = (\mu^1, \mu^2, \ldots).$$

In this sense the space (Π, d) is the projective limit space of the projective spectrum $\{\Pi^l \mid l \in \mathbb{N}\}$ together with the continuous mappings $p^l : \Pi^{l'} \to \Pi^l$, $l' \geq l$ (see, e.g., Dugundji [22], Appendix 2).

Now let H be any subset of Π, then the images $H^l := p^l(H)$ again satisfy

$$\text{(V2)} \quad p^l(H^{l+1}) = H^l \qquad \forall\, l \in \mathbb{N}.$$

In general the set H can not be uniquely reconstructed from its images H^l. If, however, H is closed then the sets H^l, being continuous images of a compact set, are compact and hence closed, and we have

$$\text{(R)} \quad H = \bigcap_{l \in \mathbb{N}} (p^l)^{-1} H^l.$$

Conversely, given a sequence $\{H^l\}$ of closed sets $H^l \subset \Pi^l$ which satisfies the compatibility conditions (V2), then there exists a uniquely determined set $H \subset \Pi$ with $H^l = p^l(H)$ for all $l \in \mathbb{N}$. This set is given by the representation (R). Consequently we may also use the notation

$$H = (H^1, H^2, \ldots)$$

for any closed set $H \subset \Pi$. For these sequence representations the following lemma holds:

Lemma 6.3. Let $H = (H^1, H^2, \ldots)$ be a closed subset of Π. Then H is connected if and only if this is true for all H^l, $l \in \mathbb{N}$.

Proof. If H is connected, the same is true for the sets H^l as continuous images of H. If the set H fails to be connected then it splits into two disjoint, non-empty, compact subsets B_1 and B_2. Letting $B_1^l := p^l(B_1)$ and $B_2^l := p^l(B_2)$ for all $l \in \mathbb{N}$, we obtain

$$\emptyset = B_1 \cap B_2 = [\bigcap_{l \in \mathbb{N}} (p^l)^{-1} B_1^l] \cap [\bigcap_{l \in \mathbb{N}} (p^l)^{-1} B_2^l]$$

$$= \bigcap_{l \in \mathbb{N}} (p^l)^{-1} (B_1^l \cap B_2^l).$$

Since the compact sets on the right-hand side of the equation form a monotonically decreasing sequence, the closed, non-empty sets B_1^l and B_2^l must be disjoint for some $l \in \mathbb{N}$. Thus the set H^l is not connected since $H^l = p^l(B_1 \cup B_2) = B_1^l \cup B_2^l$ .//

Remark 6.3. The compact set Π_{inv} of invariant W-measures possesses the sequence representation

$$\Pi_{inv} = (\Pi_{inv}^1, \Pi_{inv}^2, \ldots),$$

where

$$\Pi_{inv}^l = p^l(\Pi_{inv}) \qquad \forall\, l \in \mathbb{N}.$$

A W-measure $\mu \in \Pi^l$ is contained in Π_{inv}^l if and only if the following equations hold (see § 1.B, (W3)):

$$\sum_{b \in A} \mu(b_1, \ldots, b_{l-1}, b) = \sum_{b \in A} \mu(b, b_1, \ldots, b_{l-1}) \quad \forall\, (b_1, \ldots, b_{l-1}) \in A^{l-1}.$$

A W-measure $\mu = (\mu^1, \mu^2, \ldots) \in \Pi$ is invariant if and only if

$$\mu^l \in \Pi_{inv}^l \qquad \forall\, l \in \mathbb{N}.$$

A closed subset $H = (H^1, H^2, \ldots) \subset \Pi$ is a subset of Π_{inv} if and only if

$$H^l \subset \Pi_{inv}^l \qquad \forall\, l \in \mathbb{N}.$$

In the following definition we compare two W-measures in terms of their projections.

Definition 6.4. Let $\mu = (\mu^1, \mu^2, \ldots) \in \Pi$ and $\nu = (\nu^1, \nu^2, \ldots) \in \Pi$. Then the measure ν is called weakly continuous with respect to μ, written as

$$\nu \;<\!*\!<\; \mu,$$

if, for each $l \in \mathbb{N}$, the projection ν^l is absolutely continuous with respect to μ^l.

Remark 6.4.1. The W-measure ν is weakly continuous with respect to the W-measure μ if and only if each μ-null-cylinder is also a ν-null-cylinder.

Remark 6.4.2. Two W-measures are weakly equivalent (see Def. 2.1) if and only if they are mutually weakly continuous. Hence we have

$$\nu \overset{*}{\sim} \mu \iff [\nu <\!*\!< \mu \;\wedge\; \mu <\!*\!< \nu].$$

Definition 6.5. For all W-measures $\mu \in \Pi$ and all $l \in \mathbb{N}$ let

$$B^l(\mu) := \{\underline{b} \in A^l \mid \mu(\underline{b}) > 0\}.$$

Remark 6.5.1. The set $B^l(\mu)$ is the smallest measure-theoretic support of the W-measure $\mu^l = p^l(\mu)$ on $(A^l, \underline{A}^l)$.

Remark 6.5.2. For any two W-measures $\mu, \nu \in \Pi$ the following assertions hold:

$$\nu <\!*\!< \mu \iff [B^l(\nu) \subset B^l(\mu) \quad \forall\, l \in \mathbb{N}];$$

$$\nu \overset{*}{\sim} \mu \iff [B^l(\nu) = B^l(\mu) \quad \forall\, l \in \mathbb{N}].$$

6.C. Markov measures. A Markov measure $\mu \in \Pi$ of order l, $l \in \mathbb{N}_0$, permits the representation

$$\text{(W5)} \qquad \mu(\underline{b}) = \mu(b_1, \ldots, b_l) \cdot \Pi_{i=1}^{n} \frac{\mu(b_i,\ldots,b_{i+l})}{\mu(b_i,\ldots,b_{i+l-1})}$$

for all $n \in \mathbb{N}$ and all $\underline{b} = (b_1,\ldots,b_{n+l}) \in A^{n+l}$ (see § 1.B, (W5)). (If $l = 0$ then we have to let $\mu(b_i,\ldots,b_{i+l-1}) = 0$ by convention).

Remark 6.6.1. A Markov measure μ of order l also is of any larger order. The terminology "$l \geqslant$ order of μ" is intended to mean that μ may also be regarded as a Markov measure of order l.

Remark 6.6.2. If μ is a Markov measure of order l and if $\nu \in \Pi_{inv}$ then it follows from the representation (W5) that

$$\nu <\!*\!< \mu \iff B^{l+1}(\nu) \subset B^{l+1}(\mu).$$

In view of the representation (W5) a Markov measure μ of order l is uniquely determined by its values on the cylinders of order $l + 1$. On the other hand, a W-measure $\mu^l \in \Pi^l_{inv}$ may be extended to a Markov measure of order $l - 1$ by means of the equations (W5).

Definition 6.6. For each $l \in \mathbb{N}$ let

$$q^l : \Pi^l_{inv} \to \Pi_{inv}$$

be the mapping which assigns to each W-measure $\mu^l \in \Pi^l_{inv}$ by the equations (W5) the Markov measure μ of order $l - 1$ satisfying

$$\mu(\underline{b}) = \mu^l(\underline{b}) \qquad \forall\, \underline{b} \in A^l.$$

Remark 6.7.1. For each $l \in \mathbb{N}$ the mapping q^l is injective and continuous. The images under q^1 are identical with the Bernoulli measures.

Remark 6.7.2. For all $\mu^l \in \Pi^l_{inv}$ and all $l \in \mathbb{N}$, the relation $p^l(q^l(\mu^l)) = \mu^l$ holds.

Remark 6.7.3. Let $H^l \subset \Pi^l_{inv}$ and let $H := q^l(H^l)$ be the corresponding set of Markov measures of order $l - 1$. Then the preceding remarks imply that H is closed (and connected) if and only if the same is true for H^l.

In the sequel we sketch the well-known representation of a Markov measure as non-trivial convex combination of its ergodic components.

First $l \in \mathbb{N}$ and $B \subset A^{l+1}$ be a set of blocks.

Definition 6.7. A block $\underline{b} \in A^l$ is accessible from a block $\underline{a} \in A^l$ via B if there is a number $n \in \mathbb{N}$ and a sequence $(x_1, x_2, \ldots, x_{n+l}) \in A^{n+l}$ such that

$$(x_i, \ldots, x_{i+l}) \in B \qquad \forall\, i = 1, \ldots, n;$$

$$(x_1, \ldots, x_l) = \underline{a}; \qquad (x_{n+1}, \ldots, x_{n+l}) = \underline{b}.$$

A subset C of A^l is called irreducible relative to B if it is non-empty and if each block $\underline{b} \in C$ is accessible from each block $\underline{a} \in C$ via B. The maximal subsets of A^l which are irreducible relative to B are called the components of A^l relative to B.

Remark 6.8.1. The pair (A^l, B) may be construed as a directed graph with A^l as set of vertices and $\{((b_1, \ldots, b_l), (b_2, \ldots, b_{l+1})) \mid (b_1, \ldots, b_{l+1}) \in B\}$ as the set of directed edges. The blocks which are accessible from a block $\underline{a} \in A^l$ via B correspond to the vertices which are accessible from $\underline{a}$ via an oriented path containing at least one edge.

Remark 6.8.2. Each non-trivial cycle of the graph is contained in some component of A.

Remark 6.8.3. The components defined here are not the "strongly connected components" as defined by Berge [5], Section 3.2 inasmuch as a vertex does not necessarily lie within a component but certainly within a strongly connected component.

For graphs induced by invariant W-measures there is a particularly simple way of describing those vertices which belong to components:

Lemma 6.4. Let $\mu \in \Pi_{inv}$, then each block $\underline{b} \in B^l(\mu)$ is contained in some component of A^l relative to $B^{l+1}(\mu)$. If μ is ergodic then $B^l(\mu)$ is irreducible relative to $B^{l+1}(\mu)$.

Proof. Let $\mu \in \Pi_{inv}$ and $\underline{b} \in B^l(\mu)$. For the set $C := \bigcup_{i=0}^{\infty} T^{-i}[\underline{b}]$ one has

$$T^{-1}C \subset C \text{ and } \mu(T^{-1}C) = \mu(C).$$

This implies

$$\mu([\underline{b}] \cap T^{-1}C) = \mu(\underline{b}) > 0.$$

Thus there is an integer $n \in \mathbb{N}$ with $\mu([\underline{b}] \cap T^{-n}\, \underline{b}\,) > 0$ and a point

$x = (x_1, x_2, \ldots) \in [\underline{b}] \cap T^{-n}[\underline{b}]$ with $\mu(Z_{n+1}(x)) > 0$.

Now we have

$$(x_i, \ldots, x_{i+l}) \in B^{l+1}(\mu) \qquad \forall\, i = 1, \ldots, n;$$

$$(x_1, \ldots, x_l) = (x_{n+1}, \ldots, x_{n+l}) = \underline{b}.$$

Therefore $\underline{b}$ is accessible from itself via $B^{l+1}(\mu)$ and hence $\underline{b}$ is contained in some component of A^l relative to $B^{l+1}(\mu)$.

If furthermore μ is ergodic then the relations $T^{-1}C \subset C$ and $\mu(C) > 0$ imply that $\mu(C) = 1$. Then $\mu([\underline{a}] \cap T^{-1}C) > 0$ for each block $\underline{a} \in B^l(\mu)$. It then follows as above that $\underline{b}$ is accessible from $\underline{a}$ via $B^{l+1}(\mu)$. This proves the assertion since $\underline{a}$ and $\underline{b}$ are arbitrary blocks from $B^l(\mu)$.//

A Markov measure μ of order l is called irreducible if $B^l(\mu)$ is an irreducible subset of A^l relative to B^{l+1}; this is the case if and only if A^l has but one component relative to B^{l+1}. Furthermore the following lemma holds:

Lemma 6.5. A Markov measure is irreducible if and only if it is ergodic.

For a proof see, e.g., Billingsley [11], page 31.

Now let μ be a non-irreducible Markov measure of order l. Let $B_1, B_2, \ldots, B_r$, $r \in \mathbb{N}$, denote the components of A^l relative to $B^{l+1}(\mu)$. For each $j = 1, \ldots, r$ an irreducible Markov measure μ_j is defined by the equations

$$X_j := \bigcup_{\underline{b} \in B_j} [\underline{b}], \qquad \alpha_j := \mu(X_j)^{-1}$$

and

$$\mu_j(C) := \frac{1}{\alpha_j}\, \mu(C \cap X_j) \qquad \forall\, C \in \underline{X},$$

clearly possessing the representation

$$\mu = \Sigma_{j=1}^{r}\, \alpha_j \mu_j.$$

The Markov measures μ_j are the irreducible (ergodic) components of the Markov measure μ.

Remark 6.9. For each $n \geq 1$ there is at most one $j \in \{1, \ldots, r\}$ such that $\mu_j(Z_n(x))$ is positive. Hence the non-trivial convex combination $\mu = \Sigma_{j=1}^{r} \alpha_j \mu_j$ satisfies the assumptions of Theorem 2.16.

6.D. Generalizations with respect to P. The W-measure P relative to which Billingsley dimensions are computed in this chapter is supposed to be an ergodic (and thus irreducible), non-atomic Markov measure of finite order. Since Billingsley dimension and P-dimension coincide in the sequence space $X = A^{\mathbb{N}}$, the results of Chapter I permit us to determine the dimensions relative to many other W-measures. In particular the following remarks may be appropriate:

Remark 6.10.1. If P' is a non-atomic Markov measure, not necessarily irreducible, then it is a non-trivial convex combination of the form

$$P' = \Sigma_{i=1}^{r} \alpha_i P_i, \qquad r \in \mathbb{N},$$

of its components P_i which are irreducible, non-atomic Markov measures. According to Remark 6.9. the assumptions of Theorem 2.16 are satisfied and hence

$$P'\text{-dim}(M) = \max_{1 \leq i \leq r} P_i\text{-dim}(M) \qquad \forall\, M \subset X.$$

Remark 6.10.2. If P' is a non-invariant, non-atomic Markov measure with stationary transition probabilities then there always exists an invariant Markov measure P with the same transition probabilities. Now if furthermore the W-measures P and P' are weakly equivalent (in Example 2.5 simple citeria for this relation have been given) then they are equal by dimension according to Example 2.5, i.e. we have

$$P'\text{-dim}(M) = P\text{-dim}(M) \qquad \forall\, M \subset X.$$

Remark 6.10.3. If P' is the q-limit (q being the quasimetric of § 4) of a sequence $\{P_i'\}$ of non-atomic W-measure then we have by Theorem 4.1:

$$P'\text{-dim}(M) = \lim_{i \to \infty} P_i\text{-dim}(M) \qquad \forall\, M \subset X.$$

In this manner we can also exhibit Billingsley dimensions with respect to W-measures P' which are contained in the q-closure of the Markov measures discussed in the two preceding remarks.

It becomes apparent from these remarks that the knowledge of Billingsley dimensions with respect to irreducible, non-atomic Markov measures entails the knowledge of this dimension relative to a much larger class of W-measures.

6.E. Special functions. In this sub-section we summarize a few properties of special functions which shall be needed in the sequel. The verification of these properties is, in general, elementary or it can be accomplished by methods of calculus.

a) For each $l \in \mathbb{N}$, $h, \mu \in \Pi$ the quantity

$$E^l(h, \mu) := -\sum_{\underline{b} \in A^l} h(\underline{b}) \ln \mu(\underline{b}/\underline{b}')$$

has been defined in § 1.B. This function has the following properties:

1. If $l' \geq l$ and $l'' \geq l$ then $E^l(h, \mu)$ is also defined for $h \in \Pi^{l'}$ and $\mu \in \Pi^{l''}$.

2. $E^l(h, \mu) < \infty \iff B^l(h) \subset B^l(\mu)$.

3. Let $h, h' \in \Pi$ with $h <\!\!*\!\!< \mu$ and $h' <\!\!*\!\!< \mu$, then

$$E^l(\alpha h + (1-\alpha)h', \mu) = \alpha E^l(h, \mu) + (1-\alpha)E^l(h', \mu) \qquad \forall\, \alpha \in [0,1].$$

This inequality also holds if $E^l(h, \mu) = E^l(h', \mu) = \infty$.

4. For fixed $\mu \in \Pi$ the restriction of $E^l(h, \mu)$ onto the set $\{h \in \Pi \mid h <*< \mu\} \subset \Pi$ is continuous with respect to h.

5. If μ is a Markov measure of order l - 1 and h is invariant then

$$E^{l'}(h, \mu) = E^l(h, \mu) \qquad \forall\, l' \geq l.$$

In this case we also write $E(h, \mu)$ instead of $E^l(h, \mu)$.

6. $E^l(h, \mu) \geq E^l(h, h) \qquad \forall\, h, \mu \in \Pi_{inv}$.

b) Considered for a fixed $h \in \Pi_{inv}$, the function

$$E^l(h) := E^l(h, h)$$

is the conditional entropy of the decomposition Z_l given Z_{l-1} relative to the probability h (see Billingsley [11], page 77). Considered as a function of $h \in \Pi_{inv}$, the entropy satisfies the following conditions:

1. $E^l(h) \leq \ln|A| \qquad \forall\, h \in \Pi_{inv}$.

2. $E^l(h)$ is continuous with respect to h.

3. For all pairs $h, h' \in \Pi_{inv}$,

$$E^l(\alpha h + (1-\alpha)h') \geq \alpha E^l(h) + (1-\alpha)E^l(h') \qquad \forall\, \alpha \in [0,1].$$

4. $E^{l+1}(h) \leq E^l(h) \qquad \forall\, l \in \mathbb{N} \qquad \forall\, h \in \Pi_{inv}$.

c) As we remarked already in § 1.B, the entropy E(h) of an invariant W-measure $h \in \Pi_{inv}$ satisfies the following equation:

1. $E(h) = \lim\limits_{l\to\infty} E^l(h)$.

Furthermore we shall need the following properties:

2. As limit of a monotonically decreasing sequence of continuous functions, E(h) is upper semicontinuous. Hence we have:

$$\forall\, \varepsilon > 0 \quad \forall\, h \in \Pi_{inv} \quad \exists\, \delta > 0 \quad \forall\, h' \in \Pi_{inv} :$$

$$d(h, h') < \delta \Rightarrow E(h') \leq E(h) + \varepsilon.$$

3. For all $h, h' \in \Pi_{inv}$:

$$E(\alpha h + (1-\alpha)h') = \alpha E(h) + (1-\alpha)E(h') \qquad \forall\, \alpha \in [0,1].$$

4. The function E(h) assumes its maximum on closed (and hence on compact) subsets of Π_{inv} (this follows from 2.). On a closed and convex subset of Π_{inv} the maximum is assumed at some extremal point (this follows from 2. and 3. with Theorem 25.9 of Choquet [17]).

d) Let $\mu \in \Pi_{inv}$ be a fixed Markov measure. Then the function

$$\gamma(h, \mu) := \frac{E(h)}{E(h,\mu)} \qquad \forall\, h \in \Pi_{inv}$$

could be considered as an entropy of h with μ-correction. It has the following properties:

1. $0 \leq \gamma(h, \mu) \leq 1 \qquad \forall\, h \in \Pi_{inv}; \qquad \gamma(\mu, \mu) = 1.$

2. The function $\gamma(h, \mu)$ is upper semicontinuous on Π_{inv} with respect to h.

3. For all $h, h' \in \Pi_{inv}$ we have

   $\gamma(\alpha h + (1-\alpha)h', \mu) \geq \min\{\gamma(h, \mu), \gamma(h', \mu)\} \qquad \forall\, \alpha \in [0,1].$

4. For each constant $c \geq 0$, the set $\{h \in \Pi_{inv} \mid \gamma(h, \mu) \geq c\}$ is convex and closed in Π_{inv}.

5. For each subset $H \subset \Pi_{inv}$ one has

$$\begin{aligned}\inf\{\gamma(h, \mu) \mid h \in H\} &= \inf\{\gamma(h, \mu) \mid h \in \overline{H}\} = \inf\{\gamma(h, \mu) \mid h \in \langle H\rangle\}\\ &= \inf\{\gamma(h, \mu) \mid h \in \langle\overline{H}\rangle\}.\end{aligned}$$

(The symbol $\langle H\rangle$ denotes the convex hull of H).

§ 7. The Billingsley dimension of the smallest saturated sets

In this section we shall determine the Billingsley dimension of the smallest saturated subsets of the sequence space X (see Def. 6.3). The Hausdorff dimension of the corresponding sets of real numbers was computed by Colebrook [20]. As we have already remarked in the Introduction, Colebrook's argument can not be carried over to the general case of the dimensions considered here.

In this section let $H \in \mathbf{H}$ be arbitrary but fixed.

First we shall give an upper bound for the Billingsley dimension of the smallest saturated set M_H in terms of a μ-P-dimension which in turn shall be estimated by means of the infimum of entropies with P-correction. In order to accomplish this, let μ_0 be a W-measure satisfying

(M) $\quad \mu_0 \underset{\dim}{\leq} \mu \qquad$ for all Markov measures μ on $A^{\mathbb{N}}$,

its existence being guaranteed by Theorem 4.5. Furthermore, let

$$c := c(H) := \inf\{\gamma(h, P) \mid h \in H\}$$

(for the partial ordering $\underset{\dim}{\leq}$ and for the function $\gamma(h, P)$ see Def. 2.6 and § 6.E,d, respectively).

Theorem 7.1. $P\text{-}\dim(M_H) \leqslant \mu_0\text{-}P\text{-}\dim(M_H) \leqslant c$.

Proof. The first inequality of the assertion follows from the definition of the P-dimension (see Def. 2.4). In order to establish the second inequality we distinguish two cases. Let P be of order $k - 1$.

Case 1: There exists an $h \in H$ and a block $\underline{b} \in A^k$ satisfying

$$P(\underline{b}) = 0 \text{ and } h(\underline{b}) > 0.$$

Then $E(h, P) = \infty$ by definition, hence $c = 0$. On the other hand, for each $x \in M$ there exists by the definition of the set M_H an integer $n \in \mathbb{N}$ such that

$$|h_n(x, \underline{b}) - h(\underline{b})| < \frac{1}{2} h(\underline{b}).$$

Thus $h_n(x, \underline{b}) > 0$, i.e. the point x is contained in some P-null-cylinder. Therefore M is contained in the union of all P-null-cylinders, and hence we have (see Theorem 2.6 and Remark 2.2.3):

$$0 = P\text{-}\dim(M_H) = \mu_0\text{-}P\text{-}\dim(M_H).$$

This proves the assertion in the case under consideration.

Case 2: Suppose $h <\!*\!< P$ for all $h \in H$. According to Remark 6.6.2 this is the alternative to Case 1.

Let $\varepsilon > 0$ be chosen arbitrarily. For this ε there exists an $h \in H$ satisfying

$$\gamma(h, P) < c + \varepsilon;$$

hence there exists an integer $l \in \mathbb{N}$ with $l \geqslant k$ such that

$$\frac{E^l(h)}{E^l(h,P)} < c + 2\varepsilon .$$

Now if we choose $\alpha \in (0,1)$ sufficiently small then there exists a uniquely determined Markov measure μ of order $l - 1$ (see § 6.C) with

$$\mu(\underline{b}) = (1-\alpha) \cdot h(\underline{b}) + \alpha(P(\underline{b}) \quad \forall\, \underline{b} \in A^l$$

which satisfies

$$\frac{E(h,\mu)}{E(h,P)} = \frac{E^l(h,\mu)}{E^l(h,P)} < \frac{E^l(h,h)}{E^l(h,P)} + \varepsilon,$$

as a continuity argument will show. For each $x \in M_H$ there exists a strongly monotonic sequence $\{n_i\}$ of natural numbers satisfying

$$\lim_{i\to\infty} h_{n_i}(x) = h \qquad \text{in } (\Pi, d)$$

since h is a distribution measure of x. Now if $P(Z_n(x)) > 0$ for all $n \in \mathbb{N}$ then we also have $\mu(Z_n(x)) > 0$ for all $n \in \mathbb{N}$ by the definition of μ; therefore (see § 1,B, (W7))

$$\lim_{i\to\infty} \frac{1}{n_i} \ln P(Z_{n_i}(x)) = -E(h, P)$$

and

$$\lim_{i\to\infty} \frac{1}{n_i} \ln \mu(Z_{n_i}(x)) = -E(h, \mu).$$

In view of $h <\!*\!< P$ one has $E(h, P) > 0$ (see § 6.E, a)); consequently

$$\frac{\mu_o}{P}(x) \leqslant \frac{\mu}{P}(x) \leqslant \frac{E(h,\mu)}{E(h,P)} < c + 3\varepsilon .$$

If $P(Z_n(x)) = 0$ for some $n \in \mathbb{N}$, then $\frac{\mu_o}{P}(x) = 0 < c + 3\varepsilon$. Thus, since $\varepsilon > 0$ was arbitrary, it follows that

$$\frac{\mu_o}{P}(x) \leqslant c \qquad \forall\, x \in M_H,$$

which is equivalent to μ_o-P-dim$(M_H) \leqslant c$.//

The following theorem is preparatory for the proof of inequality

$$c \leqslant \text{P-dim}(M_H).$$

It involves an additional condition in terms of a further Markov measure ν. This condition would not be needed for the proof of the inequality itself. However, the condition will enable us in § 9.C to compute also the Billingsley dimensions of subsets of the smallest saturated sets which are defined by disallowing certain states or digits or blocks in the sequence $x = (x_1, x_2, \ldots) \in M_H$.

Theorem 7.2. Let $\nu \in \Pi$ be an ergodic Markov measure and suppose

(1) $h <\!*\!< \nu <\!*\!< P \qquad \forall\, h \in H.$

Then there exists a W-measure μ on $(X, \underline{X})$ satisfying

(2) $\mu(M_H) = 1;$

(3) $\frac{\mu}{P}(x) \geqslant c \qquad \forall\, [\mu]\, x \in X;$

(4) $\mu <\!*\!< \nu.$

The proof of Theorem 7.2 shall be preceded by a sketch of the argument and by three lemmas.

Sketch of the proof. We start with a sequence $\{h_j\}$ of distribution measures belonging to H which have each point of H as limit point and whose consecutive distances, measured by the metric d, tend to zero. After having used $h_1, h_2, \ldots, h_{j-1}$ and having defined a measure μ_{j-1} on the cylinders of order n_{j-1}, we are searching for some ergodic Markov measure which is close to h_j. We then "extend" the measure μ_{j-1} by "appending" the transition probabilities of this Markov measure a sufficiently large number of times such that, by the Individual Ergodic Theorem, "most" of the relative frequencies $h_n(x)$ are close to the ergodic Markov measure and hence close to h_j. At the same time, the Shannon- McMillan-Breiman theorem yields a proposition on $\frac{\mu}{p}(x)$ for "most" x. Then we repeat this procedure for h_{j+1} instead of h_j, etc.. Thus we obtain a W-measure μ which possesses, over a long stretch of indices, the transition probabilities of some ergodic W-measure which is close to h_j for a while, then close to h_{j+1}, etc.. This construction has the property that, for almost all x, the relative frequency $h_n(x)$ is moving from h_j to h_{j+1}, then to h_{j+2}, etc. within a margin which becomes more and more narrow. This method of proof might therefore be compared to a "semi-deterministic walk of the relative frequencies $h_n(x)$ within Π".

The following lemma guarantees the existence of a suitable sequence $\{h_j\}$ in H.

Lemma 7.1. There is a sequence $\{h_j\}$ in Π with the following properties:

(1) $h_j \in H \qquad \forall\, j \in \mathbb{N}$;

(2) $\lim_{j\to\infty} d(h_j, h_{j+1}) = 0$;

(3) H is the set of limit points of the sequence $\{h_j\}$.

Proof. We only use the fact that H is a compact, connected subset of the metric space (Π, d). For each $n \in \mathbb{N}$ there are finitely many points $y_1^n, y_2^n, \ldots, y_{r(n)}^n \in H$ satisfying

$$H \subset \bigcup_{i=1}^{r(n)} U^d(y_i^n, \tfrac{1}{n}).$$

We now consider the points y_i^n as knots of a graph, joining y_i^n to y_j^n whenever $H \cap U^d(y_i^n, \frac{1}{n}) \cap U^d(y_j^n, \frac{1}{n})$ is non-empty. Since H is topologically connected, the corresponding graph must also be connected. Hence there exists a sequence $\{h_j\}$ beginning with all y_i^1, followed by y_i^2 etc., allowing repetitions, such that

$$d(h_j, h_{j+1}) < \frac{2}{n} \quad \text{for all } h_j \in \{h_1^n, \ldots, h_{r(n)}^n\} .$$

The sequence h_j has the stated properties.//

The following additional properties of the metric d on Π and of the relative frequencies shall be needed for the proof of Theorem 7.2:

Lemma 7.2. a) For any $h_1, h_2, h_1', h_2' \in \Pi$ and $\alpha \in [0,1]$ the following inequality holds:

$$d(\alpha h_1 + (1-\alpha)h_1', \alpha h_2 + (1-\alpha)h_2') \leq \alpha d(h_1, h_2) + (1-\alpha) \cdot d(h_1', h_2').$$

b) For any $n \in \mathbb{N}$ and $x, y \in X$ with $Z_n(x) = Z_n(y)$, the inequality

$$d(h_n(x), h_n(y)) \leq \frac{2}{n} \quad \text{holds.}$$

c) Let $x \in X$ and $n, m \in \mathbb{N}$ with $n > m$. Then

$$h_n(x) = \frac{m}{n} h_m(x) + \frac{n-m}{n} h_{n-m}(T^m x) .$$

Proof. Part a) is an immediate consequence of the corresponding inequality for real numbers under the metric of the absolute value.

b) The equation $Z_n(x) = Z_n(y)$ implies

$$|h_n(x, \underline{b}) - h_n(y, \underline{b})| \leq \frac{1-1}{n} \qquad \forall\, \underline{b} \in A^1 \qquad \forall\, 1 \in \mathbb{N},$$

which yields the assertion by taking sums.

c) For each set $B \subset X$ and its characteristic function χ_B we have

$$\Sigma_{i=0}^{n-1} \chi_B(T^i x) = \Sigma_{i=0}^{m-1} \chi_B(T^i x) + \Sigma_{i=m}^{n-1} \chi_B(T^i x),$$

from which the assertion follows by elementary operations.//

The theoretical background for "appending" the transition probabilities of a W-measure to a given one is as follows:

Let $\mu/\underline{A}^n$ be the conditional probability of the W-measure μ given the σ-algebra $\underline{A}^n$. For μ-almost all $x \in X$ we then have

$$\mu/\underline{A}^n(M)(x) = \frac{\mu(M \cap Z_n(x))}{\mu(Z_n(x))} \qquad \forall\, M \in \underline{X}.$$

This means that, except for μ-null sets, $\mu/\underline{A}^n$ is a Markov kernel of $(X, \underline{X})$ relative to $(X, \underline{X})$ (see Bauer [3], § 56). This kernel transforms a W-measure ν on $(X, \underline{X})$, whose restriction $\nu^n = p^n(\nu)$ onto $\underline{A}^n$ is absolutely continuous with respect to $\mu^n = p^n(\mu)$, into the W-measure $\nu*\mu/\underline{A}^n$ (this notation differs from that which we used in § 4.B) such that

$$\begin{aligned}\nu*\mu/\underline{A}^n(M) &= \int \mu/\underline{A}^n(M)\, d\nu \\ &= \sum_{\underline{b}\in A^n} \frac{\mu(M \cap [\underline{b}])}{\mu(\underline{b})}\, \nu(\underline{b}) \qquad \forall\, M \in \underline{X}.\end{aligned}$$

If $\underline{a} = (a_1, \ldots, a_n) \in A^n$ and $\underline{b} = (b_1, \ldots, b_m) \in A^m$ then we define

$$(\underline{a}, \underline{b}) := (a_1, \ldots, a_n, b_1, \ldots, b_m) \in A^{n+m}.$$

In this notation we have

$$\nu*\mu/\underline{A}^n(\underline{a}, \underline{b}) = \nu(\underline{a}) \frac{\mu(\underline{a},\underline{b})}{\mu(\underline{a})} \qquad \forall\, \underline{a} \in A^n \qquad \forall\, \underline{b} \in A^m \qquad \forall\, m \in \mathbb{N}_0.$$

The following lemma summarizes those properties of this construction which are needed for the proof of Theorem 7.2.

Lemma 7.3. Let ν be a W-measure on $(X, \underline{X})$ and let μ be a Markov measure of order 1 which is weakly equivalent to ν. Furthermore, let

$$c(\mu) := \min \{\mu(\underline{b}) \mid \underline{b} \in B^1(\mu)\}.$$

Then the following propositions are true for m, n $\in \mathbb{N}$ with $n \geq 1$:

(1) $\nu*\mu/\underline{A}^n \overset{*}{\sim} \nu$.

(2) $\nu*\mu/\underline{A}^n(B) = \nu(B) \qquad \forall B \in \underline{A}^n$.

(3) $\nu*\mu/\underline{A}^n(\underline{a}, \underline{b}, \underline{c}) = \nu(\underline{a}, \underline{b}) \cdot \frac{\mu(\underline{b},\underline{c})}{\mu(\underline{b})} \quad \forall \underline{a} \in A^{n-1} \quad \forall \underline{b} \in A^1 \quad \forall \underline{c} \in A^m$.

(4) $\nu*\mu/\underline{A}^n(\underline{a}, \underline{c}) \leq \nu(\underline{a}) \frac{\mu(\underline{c})}{c(\mu)} \qquad \forall \underline{a} \in A^n \quad \forall \underline{c} \in A^m$.

(5) $\nu*\mu/\underline{A}^n(B \cap T^{-n}M) \geq \nu(B) \cdot [1-\frac{1-\mu(M)}{c(\mu)}] \qquad \forall B \in \underline{A}^n \quad \forall M \in \underline{X}$.

Proof. The reader should consider Definition 2.1 of weak equivalence, written as $\overset{*}{\sim}$, of two W-measures. The proof of (1) is contained in Example 2.5. Assertion (2) is true in general and follows from the definition given above. Equation (3) stems from the Markov property of the invariant W-measure μ. By estimating both numerator and denominator, (3) yields the inequality (4). First we have, for any block $\underline{b} \in A^n$ and any set $M \in \underline{X}$, the relation

$$\nu*\mu/\underline{A}^n([\underline{b}] \cap T^{-n}M) = \nu(\underline{b}) \frac{\mu([\underline{b}] \cap T^{-n}M)}{\mu(\underline{b})} \geq \nu(\underline{b}) \cdot [1 - \frac{1-\mu(M)}{c(\mu)}] ,$$

which furnishes the general inequality (5) by summing on all $\underline{b} \in B$.//

Proof of Theorem 7.2. 1. The sequence $\{\mu_j\}$. Suppose the conditions of the theorem are satisfied. By Lemma 7.1 there exists a sequence $\{h_j\}$ in H such that

(5) $\lim_{j\to\infty} d(h_j, h_{j+1}) = 0$

and

(6) H is the set of limit points of the sequence $\{h_j\}$ in Π.

Let $\{\delta_j\}$ be an arbitrary monotonic null sequence of positive numbers. For each $j \in \mathbb{N}$ we choose $l_j \in \mathbb{N}$ sufficiently large such that

(7) $l_j \geq$ order of P and $l_j \geq$ order of ν;

(8) $2^{-l_j} < \frac{1}{2} \delta_j$.

In view of (8) and the fact that the entropy function is an affine mapping we can choose, for any $j \in \mathbb{N}$, an $\alpha_j \in (0,1)$ which is small enough such that the Markov measure μ_j of order l_j with

(9) $\mu_j(\underline{b}) = (1-\alpha_j) \cdot h_j(\underline{b}) + \alpha_j\nu(\underline{b}) \qquad \forall \underline{b} \in A^{l_j+1}$

has the following properties:

(10) $d(\mu_j, h_j) < \delta_j$;

(11) $E(\mu_j) > E(h_j) - \delta_j$

(for the existence of μ_j see § 6.C). It follows from the assumption (1), $h_j \in H$ and

Definition (9) in connection with (7) that the Markov measures μ_j and ν are weakly equivalent. Hence μ_j is, like ν, irreducible and thus ergodic (see Lemma 6.4). Hence the Individual Ergodic Theorem (see § 1.B) implies that

$$\lim_{n\to\infty} d(h_n(x), \mu_j) = 0 \qquad \forall\, [\mu_j]\ x \in X,$$

and by the Shannon-McMillan-Breiman Theorem (see § 1.B) we have

$$\lim_{n\to\infty} \frac{1}{n} \ln \mu_j(Z_n(x)) = -E(\mu_j) \qquad \forall\, [\mu_j]\ x \in X.$$

Now let, for any $j \in \mathbb{N}$, the constant $c(\mu_j)$ be defined as in Lemma 7.3 and let

(12) $\varepsilon_j := 2^{-j}\, c(\mu_j)$,

then there exists, in view of the theorems stated above, a bound $m_j \in \mathbb{N}$ and a measurable set $M_j \subset X$ such that the following inequalities hold:

(13) $\mu_j(M_j) > 1 - \varepsilon_j$;

(14) $d(h_n(x), \mu_j) < \varepsilon_j \qquad \forall\, x \in M_j \qquad \forall\, n \geq m_j$;

(15) $\frac{-1}{n} \ln \mu_j(Z_n(x)) > E(\mu_j) - \delta_j \qquad \forall\, x \in M_j \qquad \forall\, n \geq m_j$.

The quantities m_j may be replaced by larger ones in such a way that they satisfy the additional conditions

(B1) $\frac{2}{n} < \delta_j, \quad \frac{1}{n}|\ln c(\mu_j)| < \delta_j \qquad \forall\, n \geq m_j$.

Now we determine successively natural numbers $n_1, n_2, \ldots$ such that, letting

(16) $r_1 := 0, \quad r_{j+1} := r_j + n_j \qquad \forall\, j \in \mathbb{N}$,

the sequence $\{n_j\}$ satisfies the conditions

(B2) $\max\,\{r_j, m_{j+1}, 2\} < \delta_j n_j \qquad \forall\, j \in \mathbb{N}$ and

(B3) $(r_j + m_{j+1}) \cdot E(\mu_j) < \delta_j n_j \qquad \forall\, j \in \mathbb{N}$.

2. The W-measure μ. Each of the Markov measures μ_j is weakly equivalent to ν; hence they are all mutually weakly equivalent. Thus, letting

(17) $\mu'_1 := \mu_1$;

$\mu'_{j+1} := \mu'_j * \mu_{j+1}/\underline{A}^{r_{j+1}} \qquad j = 1, 2, \ldots ,$

each W-measure of the sequence $\{\mu'_j\}$ is also weakly equivalent to ν. Since the sequence $\{r_j\}$ is strictly increasing and since

$$\mu'_i\big|_{\underline{A}^{r_{j+1}}} = \mu'_j\big|_{\underline{A}^{r_{j+1}}} \qquad i \geq j \geq 1,$$

the sequence $\{\mu_j'\}$ of W-measures converges in (Π, d) to a W-measure $\mu \in \Pi$. Inasmuch as

(18) $\quad \mu(B) = \mu_j'(B) \qquad \forall\, B \in \underline{A}^{r_{j+1}} \qquad \forall\, j \in \mathbb{N},$

the W-measures μ and ν are also weakly equivalent. In particular, the proposition (4) holds.

3. The set $M \subset M_H$. We introduce the following sets:

$$M_j' := \bigcup_{x \in M_j} Z_{n_j}(x) \qquad \forall\, j \in \mathbb{N}$$

$$M' := \bigcup_{m=1}^{\infty} \bigcap_{j=m}^{\infty} T^{-r_j}(M_j') \qquad \text{and}$$

$$M := \{x \in M' \mid \mu(Z_n(x)) > 0 \quad \forall\, n \in \mathbb{N}\}.$$

Clearly all these sets are measurable. Since the sets M_j' belong to the σ-algebras $\underline{A}^{n_j}$, the intersection $\bigcap_{j=m}^{l-1} T^{-r_j}(M_j')$ belongs to the σ-algebra $\underline{A}^{r_l}$ for all $l, m \in \mathbb{N}$ with $l > m \geqslant 1$. Consequently, in view of Lemma 7.3. (5) as well as (18) and (17) the expression

$$\mu\Big(\bigcap_{j=m}^{l} T^{-r_j}(M_j')\Big) = \mu_{l-1}' * \mu_l/\underline{A}^{r_l}\Big(\bigcap_{j=m}^{l} T^{-r_j}(M_j')\Big)$$

is bounded from below by

$$\mu\Big(\bigcap_{j=m}^{l-1} T^{-r_j}(M_j')\Big) \cdot \Big[1 - \frac{1}{c(\mu_l)}\,(1 - \mu_l(M_l'))\Big] \geqslant \;\ldots\; \geqslant \; 1 - 2^{-m+1},$$

where the last inequalities have been obtained by splitting the expressions into products and substituting (13) and (12) for $\mu_j(M_j') \geqslant \mu_j(M_j)$. Letting l and then m tend to infinity, monotone convergence establishes the equation

(19) $\quad \mu(M) = \mu(M') = 1.$

In the sequel we still have to show

(20) $\quad H(x) = H \qquad \forall\, x \in M \qquad$ and

(21) $\quad \frac{\mu}{p}(x) \geqslant c = c(H) \qquad \forall\, x \in M\,,$

after which the relation $M \subset M_H$ follows from (20), and the assertions (2) and (3) of the theorem are implied by (19) and (21).

For the remainder of the proof let $x \in M$ be arbitrary but fixed. Then there exists an integer $m \in \mathbb{N}$ such that

$$x \in T^{-r_j}(M_j') \qquad \forall\, j \geq m.$$

Furthermore, by the definition of the set M_j there exists, for each

(22) $x^j := T^{r_j}(x) \in M_j', \qquad j \geq m,$

an element

(23) $y^j \in Z_{n_j}(x^j) \cap M_j, \qquad j \geq m.$

In the investigation of the asymptotic behaviour of the sequences $\{h_n(x)\}$ and $\{Z_n(x)\}$ we shall distinguish two cases for all $n \in \mathbb{N}$ with $n \geq r_{m+1}$:

Case I: $\quad r_{j+1} \leq n < r_{j+1} + m_{j+1}, \qquad j \geq m;$

Case II: $\quad r_j + m_j \leq n < r_{j+1}, \qquad j \geq m+1.$

In order to understand the subsequent arguments the following interpretation may perhaps be helpful: We may consider n as a discrete time-parameter. From the time r_j up to the time $r_{j+1} - 1$ the transition probabilities of the Markov measure μ_j are valid which bring the system "close" to μ_j. However, sufficient closeness is accomplished only from the time $r_j + m_j$ onwards. Therefore the time intervals of Case I have to be distinguished from those of Case II.

4. On the sequence $\{h_n(x)\}$. In Case I we may write, using Lemma 7.2.c):

$$h_n(x) = \frac{r_j}{n} h_{r_j}(x) + \frac{n_j}{n} h_{n_j}(x^j) + \frac{n-r_{j+1}}{n} h_{n-r_{j+1}}(x^{j+1}).$$

We now split h_j accordingly, thus obtaining by Lemma 7.2.a)

$$d(h_n(x), h_j) \leq \frac{r_j}{n} + \frac{n_j}{n} d(h_{n_j}(x^j), h_j) + \frac{n-r_{j+1}}{n}$$

$$\leq 2\delta_j + d(h_{n_j}(x^j), h_{n_j}(y^j)) + d(h_{n_j}(y^j), h_j)$$

$$\leq 2\delta_j + \frac{2}{n_j} + 2\delta_j;$$

hence

(24.I) $\quad d(h_n(x), h_j) \leq 5\delta_j.$

In order to obtain this estimate we have used Condition (B2), then (23) together with Lemma 7.2.b) and (B2) as well as (10) and (14).

Analogously, we write in Case II

$$h_n(x) = \frac{r_j}{n} h_{r_j}(x) + \frac{n-r_j}{n} h_{n-r_j}(x^j),$$

thus obtaining

$$(24.\mathrm{II}) \quad d(h_n(x), \frac{r_j}{n} h_{j-1} + \frac{n-r_j}{n} h_j)$$

$$\leq \frac{r_j}{n} d(h_{r_j}(x), h_{j-1}) + \frac{n-r_j}{n} [d(h_{n-r_j}(x^j), h_{n-r_j}(y^j)) + d(h_{n-r_j}(y^j), h_j)]$$

$$\leq 5\delta_j,$$

where the first term has been estimated by (24.I) since $j \geq m + 1$; the second and third term have been handled as above, using (B1) instead of (B2).

It follows from (5), (24.I) and (24.II) that the sequences $\{h_j\}$ and $\{h_n(x)\}$ have the same limit points in Π.Therefore, in view of (6) we have $H(x) = H$, which establishes the assertion (20).

5. On the sequence $\{\frac{-1}{n} \ln \mu(Z_n(x))\}$. Using Lemma 7.3.(4) together with the relations (17) and (18), we obtain in Case I the relations

$$\mu(Z_n(x)) \leq \mu(Z_{r_{j+1}}(x)) \leq \mu(Z_{r_j}(x)) \frac{\mu_j(Z_{n_j}(x^j))}{c(\mu_j)} \leq \frac{\mu_j(Z_{n_j}(x^j))}{c(\mu_j)} ;$$

$$(25.\mathrm{I}) \quad \frac{-1}{n} \ln \mu(Z_n(x)) \geq \frac{n_j}{n} \cdot \frac{-1}{n_j} \ln \mu_j(Z_{n_j}(x^j)) - \frac{1}{n}|\ln c(\mu_j)|$$

$$\geq \frac{n_j}{n} (E(\mu_j) - \delta_j) - \delta_j \qquad \text{(by (23), (15), (B1))}$$

$$\geq E(h_j) - 4\delta_j \qquad \text{(by (11), (B2))},$$

and in Case II the inequalities

$$\mu(z_n(x)) \leq \mu(Z_{r_j}(x)) \frac{1}{c(\mu_j)} \mu_j(Z_{n_j}(x^j));$$

$$(25.\mathrm{II}) \quad \frac{-1}{n} \ln \mu(Z_n(x)) \geq \frac{r_j}{n} \cdot \frac{-1}{r_j} \ln \mu(Z_{r_j}(x))$$

$$+ \frac{n-r_j}{n} \cdot \frac{-1}{n-r_j} \ln \mu_j(Z_{n-r_j}(x^j)) + \frac{-1}{n}|\ln c(\mu_j)|$$

$$\geq \frac{r_j}{n} (E(h_{j-1}) - 4\delta_{j-1}) \quad \text{(by (25.I) since } j \geq m + 1)$$

$$+ \frac{n-r_j}{n} (E(h_j) - 4\delta_j) \quad \text{(as in (25.I))}.$$

6. On the sequence $\{\frac{-1}{n}\ln P(Z_n(x))\}$. By the construction of the set M we have $\mu(Z_n(x)) > 0$ for all $n \in \mathbb{N}$. Therefore, in view of $\mu <\!\!*\!\!< \nu <\!\!*\!\!< P$, it follows that also $P(Z_n(x)) > 0$ for all $n \in \mathbb{N}$, and hence $h_n(x)$ is always contained in the compact set

$$\Theta := \{h \in \Pi \mid h <\!\!*\!\!< P\}.$$

Now $E^{l+1}(h, P)$ is an affine and continuous function of h, hence uniformly continuous on the set Θ. In view of (24.I) and (24.II) where $\lim_{j\to\infty} \delta_j = 0$, using the inequality

$$\frac{-1}{n}\ln P(Z_n(x)) \leq \frac{-1}{n}\ln P(Z_{n+l}(x)) = \frac{-1}{n}\ln P(Z_l(x)) + E^{l+1}(h_n(x), P),$$

where l denotes the order of the Markov measure P (see § 1.B, (W7)), there exists a null sequence $\{t_j\}$ of positive real numbers satisfying

$$(26.\text{I}) \quad \frac{-1}{n}\ln P(Z_n(x)) \leq E^{l+1}(h_j, P) + t_j = E(h_j, P) + t_j$$

and

$$(26.\text{II}) \quad \frac{-1}{n}\ln P(Z_n(x)) \leq \frac{r_j}{n}(E(h_{j-1}, P) + t_{j-1}) + \frac{n-r_j}{n}(E(h_j, P) + t_j)$$

in Case I and Case II, respectively.

7. Proof of (21). Using (25.I) and (26.I) in Case I and (25.II) and (26.II) in Case II we obtain by forming quotients the inequality

$$\frac{\ln \mu(Z_n(x))}{\ln P(Z_n(x))} \geq \min \left\{\frac{E(h_i) - 4\delta_i}{E(h_i,P) + t_i} \;\middle|\; i = j-1, j\right\}.$$

Furthermore, if we note that j increases with n, that the sequences $\{\delta_j\}$ and $\{t_j\}$ converge to 0 and that the sequence $\{\frac{1}{E(h_j,P)}\}$ remains bounded, we find by comparing the lower limits of both sides that

$$\frac{\mu}{P}(x) \geq \liminf_{j\to\infty} \frac{E(h_j)}{E(h_j,P)} \geq \inf\left\{\frac{E(h)}{E(h,P)} \;\middle|\; h \in H\right\} = c.$$

This shows that (21) is also true for any point $x \in M$, which completes the proof of Theorem 7.2.//

At the moment we are interested in the following special case of Theorem 7.2:

<u>Theorem 7.3.</u> There exists a W-measure μ on $(X, \underline{X})$ such that

(1) $\mu(M_H) = 1$;

(2) $\frac{\mu}{P}(x) \geq c \qquad \forall\, [\mu]\; x \in X.$

<u>Proof.</u> If all distribution measures $h \in H$ are weakly continuous with respect to P then the conditions of Theorem 7.2 are satisfied with $\nu := P$, the assertions then containing those of Theorem 7.3. Otherwise we have (see the proof of Theorem 7.1)

$P\text{-dim}(M_H) = 0 = c.$

But the conditions of Theorem 7.2 are always satisfied for the measure P_g (see Remark 3.3) of equidistribution instead of P, letting $\nu := P_g$. The measure μ of this theorem then satisfies assertion (1) of Theorem 7.3, and because of

$$0 \leq \mu\text{-ess.sup}\, \tfrac{\mu}{P}(x) \leq P\text{-dim}(M_H) = 0$$

(see Theorem 2.12) it also satisfies Condition (2).//

After these preparations we are now able to determine the Billingsley dimension of the smallest saturated set.

Theorem 7.4. The equation

$$P\text{-dim}(M_H) = \mu_0\text{-}P\text{-dim}(M_H) = c = \inf_{h \in H} \gamma(h, P)$$

holds.

Proof. In view of Theorem 7.1 we only have to show

$$c \leq P\text{-dim}(M_H).$$

But by Theorem 2.12 the W-measure μ of Theorem 7.3 has the property

$$c \leq \mu\text{-ess.sup}_{x \in M_H}\, \tfrac{\mu}{P}(x) \leq P\text{-dim}(M_H),$$

which proves the theorem.//

Remark 7.1. Rewritten in the notation Remark 3.3, Colebrook [20] computes the Hausdorff dimension of the set $\phi^{-1}(M_H)$. This result, however, is contained as a special case in Theorem 7.4 since, by Theorem 3.4, one has

$$h\text{-dim}(\phi^{-1}(M_H)) = P_g\text{-dim} = \inf_{h \in H} \gamma(h, P_g) = \inf_{h \in H} \frac{E(h)}{\ln g}.$$

The last term is the form in which Colebrook stated his result. In the general case, when P is different from the measure P_g of equidistribution, Theorem 7.4 shows that the entropy E(h) has to be corrected by a factor which depends on the distribution measure h and on the Markov measure P and which happens to have the value $\frac{1}{\ln g}$, independent of h, only in the case $P = P_g$. Therefore the name "entropy of h corrected by P" for the expression $\gamma(h,P) = \frac{E(h)}{E(h,P)}$ appears to be justified.

Example 7.1. If ν is an invariant measure on the sequence space $X = A^{\mathbb{N}}$ then the one-point set $\{\nu\}$ is closed and connected in (Π, d), hence $\{\nu\} \in \mathbb{H}$, and $M_{\{\nu\}}$ is the set of ν-normal sequences $x = (x_1, x_2, \ldots) \in X$, corresponding to ν-normal real numbers. For the Billingsley dimension of the set $M_{\{\nu\}}$ we have by Theorem 7.4:

$$\text{P-dim}(M_{\{\nu\}}) = \frac{E(\nu)}{E(\nu,P)} .$$

On the other hand,

$$\nu(M_{\{\nu\}}) = \begin{cases} 1 \text{ if } \nu \text{ is ergodic} \\ 0 \text{ otherwise .} \end{cases}$$

(The former case follows from the Individual Ergodic Theorem, the latter from the representation of invariant W-measures in terms of ergodic measures by Lemma 1.2).

If ν and P are Bernoulli measures, one obtains the representation

$$\text{P-dim}(M_{\{\nu\}}) = \frac{\Sigma_{a\in A}\ \nu(a)\ \ln \nu(a)}{\Sigma_{a\in A}\ \nu(a)\ \ln P(a)}.$$

In the three following remarks we consider the proof of Theorem 7.3 which allows some modifications.

Remark 7.2. In the proof of Theorem 7.3 the sequence $\{h_j\}$ does not necessarily have to lie in H. It is only necessary that all h_j are weakly continuous with respect to ν, that H is the set of limit points of the sequence $\{h_j\}$ and that

$$\liminf_{j\to\infty} \gamma(h_j, P) \geq \inf_{h\in H} \gamma(h, P).$$

Furthermore the null sequence $\{\delta_j\}$ may converge arbitrarily fast. Therefore, letting $\nu := P$, we have the possibility of choosing, for each $j \in \mathbb{N}$, two different Markov measures μ_j^0 and μ_j^1 either of which may assume the role of μ_j. Then we can determine the integers m_j and n_j large enough such that propositions (13), (14) and (15) of the proof are valid for sets M_j^0 and M_j^1, respectively, and conditions (B1), (B2) and (B3) hold for μ_j^0 and μ_j^1. For each 0-1 - sequence $\underline{i} = (i_1, i_2, \ldots) \in \{0,1\}^{\mathbb{N}}$, we may then construct, by means of the sequence $(\mu_1^{i_1}, \mu_2^{i_2}, \ldots)$ of Markov measures, a W-measure $\mu^{\underline{i}}$ analogous to μ, and a set $M^{\underline{i}}$ constructed by means of the sequence $(M_1^{i_1}, M_2^{i_2}, \ldots)$. Then the relation

$$\text{P-dim}(M^{\underline{i}}) = \text{P-dim}(M_H) = \inf_{h\in H} \gamma(h, P) \qquad \forall\ \underline{i} \in \{0,1\}^{\mathbb{N}}$$

holds, including the assertion of Theorem 7.1. In addition we may assume the numbers δ_j to be so small that, for each 0-1 - sequence $\underline{i}$ and for each point x belonging to the corresponding set $M^{\underline{i}}$, there exists an integer $m(x) \in \mathbb{N}$ such that

$$d(h_{r_{j+1}}(x), \mu_j^{i_j}) < \frac{1}{2}\, d(\mu_j^0, \mu_j^1) \qquad \forall\ j \geq m(x).$$

Clearly, two sets $M^{\underline{i}}$ and $M^{\underline{i}'}$ are disjoint if the 0-1 - sequences $\underline{i}$ and $\underline{i}'$ differ at infinitely many places. This proves the following theorem:

Theorem 7.5. For each $H \in \mathbb{H}$ the set M_H may be split into non-denumerably many disjoint subsets M_α, $\alpha \in I$, such that

$$\text{P-dim}(M_\alpha) = \text{P-dim}(M_H) = \inf_{h \in H} \gamma(h, P) \qquad \forall \alpha \in I.$$

Remark 7.3. A further feature of the proof of Theorem 7.2 lies in the rate of convergence. Since the numbers n_j may be chosen arbitrarily large, a suitable choice of the sequence $\{h_j\}$ and the null sequence $\{\delta_j\}$ will make the convergence of the sequences $\{h_n(x)\}$ for the points $x \in M$ arbitrarily slow (a concept which would have to be defined precisely for sets H consisting of more than one point).

Remark 7.4. Finally it should be noted that in the proof of Theorem 7.2, for any two points x and y of the set M the relation

$$\lim_{n\to\infty} d(h_n(x), h_n(y)) = 0$$

holds. In general this is not true for two arbitrary points x and y of the set M_H. This again shows that a very small subset M of M_H has been constructed in the proof. It would be of interest to investigate which points $x \in X$ with

$$M(x) := \{y \in X \mid \lim_{n\to\infty} d(h_n(x), h_n(y)) = 0\}$$

have the property

$$\text{P-dim}(M(x)) = \text{P-dim}(M_{H(x)}).$$

§ 8. The Billingsley dimension of saturated sets

By means of Theorem 7.4 the following fundamental theorem on Billingsley dimensions of saturated sets can now be proved.

Theorem 8.1. a) For each saturated set $M_{\underline{H}}$ with $\underline{H} \subset \mathbb{H}$ the relation

$$\text{P-dim}(M_{\underline{H}}) = \sup_{H \in \underline{H}} \text{P-dim}(M_H) = \sup_{H \in \underline{H}} \inf_{h \in H} \gamma(h, P)$$

holds.

b) For every element α of an arbitrary set I of indices let B_α be a saturated subset of X. Then the equation

$$\text{(SUP)} \qquad \text{P-dim}\Big(\bigcup_{\alpha \in I} B_\alpha\Big) = \sup_{\alpha \in I} \text{P-dim}(B_\alpha)$$

is true.

Proof. a) Let $\underline{H} \subset \mathbb{H}$. Then one has $M_{\underline{H}} = \bigcup_{H \in \underline{H}} M_H$. By Theorem 7.4,

$$\text{P-dim}(M_H) = \mu_0\text{-P-dim}(M_H) \qquad \forall H \in \underline{H},$$

which implies by Theorem 2.7:

$$\text{P-dim}(M_{\underline{H}}) = \sup_{H \in \underline{H}} \text{P-dim}(M_H).$$

This is the first equation of the assertion. The second one is obtained by applying Theorem 7.4 once more.

b) For each $\alpha \in I$ there exists a set $\underline{H}_\alpha \subset \mathbb{H}$ satisfying $B_\alpha = M_{\underline{H}_\alpha} = \bigcup_{H \in \underline{H}_\alpha} M_H$. Letting

$\underline{H}_I := \bigcup_{\alpha \in I} \underline{H}_\alpha$, we obtain by part a):

$$\text{P-dim}(\bigcup_{\alpha \in I} B_\alpha) = \text{P-dim}(\bigcup_{H \in \underline{H}_I} M_H) = \sup_{H \in \underline{H}_I} \text{P-dim}(M_H)$$

$$= \sup_{\alpha \in I} \sup_{H \in \underline{H}_\alpha} \text{P-dim}(M_H) = \sup_{\alpha \in I} \text{P-dim}(B_\alpha),$$

which proves part b) of the theorem.//

Remark 8.1.1. It is impossible that the relation (SUP) could hold without any assumptions on the sets B_α since, e.g., the one-element sets $\{x\}$, $x \in X$, satisfy the equations

$$\sup_{x \in X} \text{P-dim}(\{x\}) = 0 \quad \text{and} \quad \text{P-dim}(\bigcup_{x \in X} \{x\}) = \text{P-dim}(X) = 1.$$

Remark 8.1.2. For the Billingsley dimension of an arbitrary subset B of X, Theorem 8.1.a) yields the upper bound

$$\text{P-dim}(B) \leqslant \sup_{x \in B} \inf_{h \in H(x)} \gamma(h, P)$$

which is useful whenever the distribution measures $h \in H(x)$ of its points are known.

Remark 8.1.3. A further justification for calling sets of the form $M_{\underline{H}}$ saturated sets is supplied by Theorem 8.1.b); these sets are so large and "saturated" within X that the Billingsley dimension of a non-denumerable union of them does not become larger than it would have to be trivially on account of the individual sets involved.

We shall also reformulate Theorem 8.1 for the Hausdorff dimension of the corresponding subsets of the interval [0,1]. In Remark 3.3 we have mapped the interval [0,1] into the space $X = A^{\mathbb{N}}$ by means of the g-adic digit expansion

$$x = \sum_{i=1}^{\infty} g^{-i} e_i(x), \quad e_i(x) \in A = \{0, 1, \ldots, g-1\}, \; g \in \mathbb{N}, \; g \geqslant 2,$$

of a number $x \in [0,1]$ in such a way that dimensions were preserved, choosing the Hausdorff dimension "h-dim" on [0,1] and the Billingsley dimension with respect to the measure P_g of equidistribution (which is the Bernoulli measure satisfying $P_g(a) = \frac{1}{g}$ for all $a \in A$) on the space $A^{\mathbb{N}}$. With each number $x \in [0,1]$ this procedure associates the sequence $\phi(x) = (e_1(x), e_2(x), \ldots) \in X$ to which the set $H(\phi(x)) \in \mathbb{H}$ of distribution measures belongs.

Definition 8.1. For every non-empty, closed and connected set $H \in \mathbb{H}$ of invariant W-measures let

$$F_H := \{x \in [0,1] \mid H(\phi(x)) = H\}.$$

For every subset $\underline{H} \subset \mathbb{H}$ let

$$F_{\underline{H}} := \{x \in [0,1] \mid H(\phi(x)) \in \underline{H}\}.$$

Sets of the form $F_{\underline{H}}$, $\underline{H} \subset \mathbb{H}$, are called (g-adically) saturated. Sets of the form F_H, $H \in \mathbb{H}$, are the smallest (g-adically) saturated subsets of the interval [0,1].

We are now able to formulate the Fundamental Theorem on the Hausdorff dimension of g-adically saturated subsets of the interval [0,1].

Theorem 8.2. a) For every g-adically saturated set $F_{\underline{H}}$ with $\underline{H} \subset \mathbb{H}$ the relation

$$\text{h-dim}(F_{\underline{H}}) = \sup_{H \in \underline{H}} \text{h-dim}(F_H) = \sup_{H \in \underline{H}} \inf_{h \in H} \frac{E(h)}{\ln g}$$

holds.

b) For every element α of a given set I of indices let $g_\alpha \in \mathbb{N}$, $g_\alpha \geq 2$, and let F_α be a g_α-adically saturated subset of the interval [0,1]. Then the equation

$$\text{(SUP)} \qquad \text{h-dim}\Big(\bigcup_{\alpha \in I} F_\alpha\Big) = \sup_{\alpha \in I} \text{h-dim}(F_\alpha)$$

is valid.

Proof. a) For every set $H \in \mathbb{H}$ we have $F_H = \phi^{-1}(M_H)$, hence also $F_{\underline{H}} = \phi^{-1}(M_{\underline{H}})$ for all $\underline{H} \subset \mathbb{H}$ and thus the assertion follows immediately from Theorem 8.1 by means of Theorem 3.4 if we replace the measure P there by P_g, keeping in mind that

$$E(h, P_g) = \ln g \qquad \forall\, h \in \Pi_{inv}.$$

b) For each $g \in \mathbb{N}$ with $g \geq 2$ let $I_g := \{\alpha \in I \mid g_\alpha = g\}$. By part a) of the theorem or by Theorem 8.1.b) it follows then that

$$\text{h-dim}\Big(\bigcup_{\alpha \in I} F_\alpha\Big) = \text{h-dim}\Big(\bigcup_{g=2}^{\infty} \bigcup_{\alpha \in I_g} F_\alpha\Big) = \sup_{\substack{g \in \mathbb{N} \\ g \geq 2}} \text{h-dim}\Big(\bigcup_{\alpha \in I_g} F_\alpha\Big)$$

$$= \sup_{\substack{g \in \mathbb{N} \\ g \geq 2}} \sup_{\alpha \in I_g} \text{h-dim}(F_\alpha) = \sup_{\alpha \in I} \text{h-dim}(F_\alpha).$$

This establishes part b) of the theorem.//

Remark 8.2. By substituting one-point sets $\underline{H} = \{H\}$, $H \in \mathbb{H}$, in Theorem 8.2.a) we obtain the Hausdorff dimension of the smallest g-adically saturated subsets of [0,1] as they have already been computed by Colebrook [20].

Before we are going to show by means of examples how Theorems 8.1 and 8.2 facilitate the computation of Billingsley and Hausdorff dimensions of saturated sets, we shall investigate the continuity of the functions P-dim(M_H) and P-dim($M_{\underline{H}}$) with respect to H and $\underline{H}$. This problem will be the subject of the following two theorems and a counter-example. The spaces $\mathbb{H}$ and $\underline{\mathbb{H}}$ have to be equipped with the metric δ and the quasimetric $\underline{\delta}$, respectively, as introduced in Definition 6.3.

Theorem 8.3. For fixed P the function P-dim(M_H) is upper semicontinuous on the metric space $(\mathbb{H}, \delta)$.

Proof. Let $H \in \mathbb{H}$ and $\varepsilon > 0$ be given. By Theorem 7.4 there exists a distribution measure $h \in H$ with $\gamma(h, P) <$ P-dim(M_H) $+ \frac{\varepsilon}{2}$. Since the function $\gamma(\cdot, P)$ is upper semicontinuous (see § 6.E,d)), there exists a number $s > 0$ such that $\gamma(h', P) < \gamma(h,P) + \frac{\varepsilon}{2}$ for all $h' \in \Pi_{inv}$ with $d(h, h') < s$.

Now if $H' \in \mathbb{H}$ is such that $\delta(H, H') < s$ then there exists a W-measure h' within the set H' satisfying $d(h, h') < s$. By Theorem 7.4 this implies

$$\text{P-dim}(M_{H'}) \leqslant \gamma(h', P) < \gamma(h, P) + \frac{\varepsilon}{2} < \text{P-dim}(M_H) + \varepsilon,$$

which proves the theorem.//

Corollary 8.3. Let $\{H_i\}$ be a sequence in $(\mathbb{H}, \delta)$ which converges to some element $H_0 \in \mathbb{H}$. Then the inequality

$$\text{P-dim}(M_{H_0}) \geqslant \limsup_{i\to\infty} \text{P-dim}(M_{H_i})$$

holds. The relation

$$\text{P-dim}(M_{H_0}) = \lim_{i\to\infty} \text{P-dim}(M_{H_i})$$

is true, e.g., if all sets H_i are subsets of H_0 since in this case one has

$$\text{P-dim}(M_{H_i}) \geqslant \text{P-dim}(M_{H_0}) \qquad \text{for all } i \in \mathbb{N}.$$

Theorem 8.4. Let $\underline{H}_0$ be a δ-closed subset of $\mathbb{H}$. Then P-dim($M_{\underline{H}}$) is, for fixed P, regarded as a function of $\underline{H}$ on the quasimetric space $(\underline{\mathbb{H}}, \underline{\delta})$, upper semicontinuous at the point $\underline{H}_0 \in \underline{\mathbb{H}}_0$.

Proof. In order to prove the assertion indirectly, let us assume that $\underline{H}_0$ is a δ-closed subset of $(\mathbb{H}, \delta)$ such that P-dim($M_{\underline{H}}$) is not upper semicontinuous at the point $\underline{H}_0$. Then there exists an $\varepsilon > 0$ and a sequence $\{\underline{H}_i\}$ in $\underline{\mathbb{H}}$ satisfying

(1) $\lim_{i\to\infty} \underline{\delta}(\underline{H}_i, \underline{H}_0) = 0$

and

$$\text{P-dim}(M_{\underline{H}_i}) > \text{P-dim}(M_{\underline{H}_o}) + \varepsilon \qquad \forall\, i \in \mathbb{N}.$$

Then there exists by Theorem 8.1 in each set $\underline{H}_i \subset \mathbb{H}$ an element $H_i \in \mathbb{H}$ such that

$$\text{P-dim}(M_{H_i}) > \text{P-dim}(M_{\underline{H}_o}) + \varepsilon \qquad \forall\, i \in \mathbb{N}.$$

Since the space $(\mathbb{H}, \delta)$ is sequentially compact (see Lemma 6.2) there exists a convergent subsequence of $\{H_i\}$ to be denoted by $\{H_i\}$ again. In view of (1) and since $\underline{H}_o$ is δ-closed, the limit $H_o \in \mathbb{H}$ of this sequence belongs to $\underline{H}_o$. Therefore

$$\text{P-dim}(M_{\underline{H}_o}) \geq \text{P-dim}(M_{H_o}) \geq \limsup_{i\to\infty} \text{P-dim}(M_{H_i}) \geq \text{P-dim}(M_{\underline{H}_o}) + \varepsilon.$$

This is a contradiction and thus the assertion follows.//

Corollary 8.4. If $\{\underline{H}_i\}$ is a monotonically decreasing sequence of closed subsets of the space $(\mathbb{H}, \delta)$ then the set $\underline{H}_o := \bigcap_{i\in\mathbb{N}} \underline{H}_i$ is a closed subset of $(\mathbb{H}, \delta)$ satisfying

$$\text{P-dim}(M_{\underline{H}_o}) = \lim_{i\to\infty} \text{P-dim}(M_{\underline{H}_i}).$$

Proof. The assertion is trivial if one of the sets $\underline{H}_i$ is empty. If all $\underline{H}_i$ are non-empty then it follows from the compactness of the space $(\mathbb{H}, \delta)$ (see Lemma 6.2, noting that metric, sequentially compact spaces are compact) that their intersection $\underline{H}_o$ is non-empty and compact, hence also closed, and it satisfies the equation

$$\lim_{i\to\infty} \underline{\delta}(\underline{H}_i, \underline{H}_o) = 0.$$

Now Theorem 8.4 implies

$$\text{P-dim}(M_{\underline{H}_o}) \geq \limsup_{i\to\infty} \text{P-dim}(M_{\underline{H}_i}).$$

But since $M_{\underline{H}_o}$ is contained in each of the sets $M_{\underline{H}_i}$, one has

$$\text{P-dim}(M_{\underline{H}_o}) \leq \text{P-dim}(M_{\underline{H}_i}) \qquad \forall\, i \in \mathbb{N}.$$

These two inequalities prove the assertion.//

Example 8.1. Scope of Theorems 8.3 and 8.4. Let $h_o \in \Pi_{inv}$ be an invariant W-measure on $(X, \underline{X})$ satisfying $\gamma(h_o, P) > 0$ (for example, let $h_o = P$). Furthermore let $x = (x_1, x_2, \ldots) \in M_{\{h\}}$, hence $x \in X$ and

$$\lim_{n\to\infty} d(h_n(x), h) = 0.$$

Now we consider the sequence $x^i \in X$ which is obtained by indefinite repetition of the finite section $(x_1, x_2, \ldots, x_i)$ of the sequence x. The relative frequencies $h_n(x^i)$ converge, for each point x^i, to an atomic, invariant Markov measure h_i of order i with $E(h_i) = 0$. Now we have

$$\lim_{i\to\infty} d(h_i, h_o) = 0 \qquad \text{and}$$

$$\gamma(h_i, P) = 0 \qquad \forall\, i \in \mathbb{N}.$$

(Therefore the entropy function can not be continuous at the point h_o since $E(h_o) > 0$). Letting $H_i := \{h_i\}$ for all $i \in \mathbb{N}_o$, we obtain

$$\lim_{i\to\infty} \delta(H_i, H_o) = 0 \qquad \text{and}$$

$$\text{P-dim}(M_{H_o}) = \gamma(h_o, P) > 0 = \gamma(h_i, P) = \text{P-dim}(M_{H_i}) \qquad \forall\, i \in \mathbb{N}.$$

Thus P-dim(M_H), considered as a function of H, can not be continuous at the point H_o. Similarly, letting $\underline{H}_i := \{H_i\} \in \mathbb{H}$ for all $i \in \mathbb{N}_o$, we obtain the result that P-dim($M_{\underline{H}}$), considered as a function of $\underline{H}$ is not continuous at the point $\underline{H}_o$ even though $\underline{H}_o$ is a non-empty, closed subset of $\mathbb{H}$.

Finally we let $\underline{H}' := \{H_1, H_2, \ldots\}$ and $\underline{H}'' := \{H_o, H_1, H_2, \ldots\}$, and we obtain for these sets the relations

$$\underline{\delta}(\underline{H}', \underline{H}'') = 0 \qquad \text{and}$$

$$\text{P-dim}(M_{\underline{H}''}) = \text{P-dim}(M_{H_o}) = \gamma(h_o, P) > 0 = \text{P-dim}(M_{\underline{H}'}).$$

This shows that in Theorem 8.4 the condition that the set $\underline{H}_o \subset \mathbb{H}$ be closed in $(\mathbb{H}, \delta)$ is indispensable.

The following two examples are intended to illustrate the way in which Theorems 8.1 and 8.2 may be applied.

<u>Example 8.2.</u> Let $H_n(x, 1)$ be the relative frequency of the digit 1 among the first n digits of the dyadic extension (g = 2) of the number $x \in [0,1]$. Furthermore let

$$u(x) := \liminf_{n\to\infty} h_n(x, 1) \qquad \forall\, x \in [0,1],$$

$$o(x) := \limsup_{n\to\infty} h_n(x, 1) \qquad \forall\, x \in [0,1] \qquad \text{and}$$

$$f(t) := \frac{t \ln t + (1-t) \ln (1-t)}{-\ln 2} \qquad \forall\, t \in [0,1]$$

with $f(0) = f(1) = 0$.

For the sets $B(\xi) := \{x \in [0,1] \mid o(x) \leq \xi\}$, $0 \leq \xi \leq \frac{1}{2}$, Besicovitch [7] showed that
$\text{h-dim}(B(\xi)) = f(\xi)$.
For the sets $C(\xi) := \{x \in [0,1] \mid u(x) < \xi\}$, $0 < \xi \leq \frac{1}{2}$, it was shown by Knichal [32] that
$\text{h-dim}(C(\xi)) = f(\xi)$.
For the sets $D(\xi) := \{x \in [0,1] \mid u(x) = \xi = o(x)$, $0 \leq \xi \leq 1$, and $E(\xi) := \{x \in [0,1] \mid u(x) \leq \xi \leq o(x)\}$, $0 \leq \xi \leq 1$ the relation

$$h\text{-dim}(D(\xi)) = h\text{-dim}(E(\xi)) = f(\xi)$$

was shown by Volkmann [44, § 8]. All these sets are dyadically saturated and may be represented as unions of saturated sets of the form

$$F(\alpha, \beta) := \{x \in [0,1] \mid u(x) = \alpha,\ o(x) = \beta\},\ 0 \leqslant \alpha \leqslant \beta \leqslant 1$$

whose Hausdorff dimension was determined by Volkmann [49] from more general results as

$$h\text{-dim}(F(\alpha, \beta)) = \min\{f(\alpha), f(\beta)\}.$$

Using this equation one can immediately determine the Hausdorff dimensions of the sets $B(\xi)$, $C(\xi)$, $D(\xi)$ and $E(\xi)$. For example we obtain

$$B(\xi) = \bigcup_{0\leqslant\alpha\leqslant\beta\leqslant\xi} F(\alpha, \beta)$$

and therefore

$$\begin{aligned} h\text{-dim}(B(\xi)) &= \sup\{h\text{-dim}(F(\alpha, \beta)) \mid 0 \leqslant \alpha \leqslant \beta \leqslant \xi\} \\ &= \sup\{\min\{f(\alpha), f(\beta)\} \mid 0 \leqslant \alpha \leqslant \beta \leqslant \xi\} \\ &= \begin{cases} f(\xi) & \text{if } 0 \leqslant \xi \leqslant \frac{1}{2}, \\ 1 & \text{if } \frac{1}{2} \leqslant \xi \leqslant 1. \end{cases} \end{aligned}$$

As far as the author knows, the Billingsley dimension of the sets considered in the following example has not yet been computed with the same generality. The example is intended to demonstrate how the computation of the Billingsley dimension of a set is reduced by means of Theorem 8.1 to solving an extremal value problem under constraints.

Example 8.3. Let $A = \{1, 2, \ldots, k\}$, $k \geqslant 2$, and let P be a Bernoulli measure on the sequence space $X = A^{\mathbb{N}}$. For arbitrary real numbers α and β with $0 \leqslant \alpha \leqslant \beta \leqslant \ln k$ we are going to determine the Billingsley dimension (relative to P) of the set

$$X_\alpha^\beta := \{x \in X \mid \alpha = \inf_{h\in H(x)} E(h),\ \beta = \sup_{h\in H(x)} E(h)\}.$$

Thus the set X_α^β is characterized by the entropy of the distribution measures of its points. The set is understood to be empty if the numbers α and β fail to satisfy the condition $0 \leqslant \alpha \leqslant \beta \leqslant \ln k$. If P is the measure P_g of equidistribution on $A^{\mathbb{N}}$ (see Remark 3.3) then it follows immediately from Theorems 8.1 and 8.2 that

$$P_g\text{-dim}(X_\alpha^\beta) = \frac{\alpha}{\ln k}.$$

If P is different from the measure P_g then we may proceed as follows: For each $x \in X_\alpha^\beta$ there exists an $h \in H(x)$ such that $E(h) = \beta$ since the entropy function is upper semi-continuous and $H(x)$ is compact. Hence we have

$$P\text{-dim}(M_{H(x)}) \leqslant \gamma(h, P).$$

In view of

(1) $X_\alpha^\beta = \bigcup_{x \in X_\alpha^\beta} M_{H(x)}$,

letting

(2) $f(t) := \max \{\gamma(h, P) \mid h \in \Pi_{inv}, E(h) = t\}$ $\forall\ t \in [0, \ln k]$,

this implies by Theorem 8.1 that

(3) $P\text{-dim}(X_\alpha^\beta) \leq f(\beta)$.

But there also exists, for each point $x \in X_\alpha^\beta$, a sequence $\{h_i\}$ of distribution measures of x satisfying $\lim_{i\to\infty} E(h_i) = \alpha$. Thus

$$P\text{-dim}(M_{H(x)}) \leq \inf_{i \in \mathbb{N}} f(E(h_i)) \leq f(\alpha),$$

and another application of Theorem 8.1 yields

(4) $P\text{-dim}(X_\alpha^\beta) \leq f(\alpha)$.

Since the function $\gamma(\cdot, P)$ is upper semicontinuous and since Π_{inv} is compact, there exist, for given α and β, W-measures $\mu_0, \mu_1 \in \Pi_{inv}$ satisfying

$$E(\mu_0) = \alpha,\ \gamma(\mu_0, P) = f(\alpha),\ E(\mu_1) = \beta,\ \gamma(\mu_1, P) = f(\beta).$$

The set $H := \{\mu_t := (1-t)\mu_0 + t\mu_1 \mid t \in [0,1]\}$, being a continuous image of the interval $[0,1]$, is a non-empty, closed and connected set of invariant W-measures, i.e. $H \in \mathbb{H}$. Because of

$$E(\mu_t) = (1-t)\ E(\mu_0) + t\ E(\mu_1) \qquad \forall\ t \in [0,1]$$

it follows that $M_H \subset X_\alpha^\beta$ and furthermore, by § 6.E, d)3. and Theorem 7.4,

(5) $P\text{-dim}(X_\alpha^\beta) \geq P\text{-dim}(M_H) = \inf \{\gamma(\mu_t, P) \mid t \in [0,1]\}$
$= \min \{\gamma(\mu_0, P), \gamma(\mu_1, P)\}$
$= \min \{f(\alpha), f(\beta)\}$.

Combining the inequalities (3), (4) and (5), we obtain the following result:

(6) $P\text{-dim}(X_\alpha^\beta) = \min \{f(\alpha), f(\beta)\}$.

In this way we have solved the problem except for determining the function $f(t)$ as defined in (2). About the measure P we have so far only used the fact that it is an ergodic Markov measure of finite order.

We shall now determine the function $f(t)$ more precisely under the assumption that P be a Bernoulli measure. For this purpose we assume without loss of generality that the elements of the state space A are numbered in such a way that, for suitable positive integers $r \leq k' \leq k$ the following conditions hold:

$$P(i) = P(1) \qquad \text{for } i = 1, 2, \ldots, r,$$

$$0 < P(i) < P(1) \qquad \text{for } i = r+1, \ldots, k',$$
$$P(i) = 0 \qquad \text{for } i = k'+1, \ldots, k.$$

The determination of the function $f(t)$ is equivalent to determining the function

$$g(t) := \min \{E^1(h, P) \mid h \in \Pi_{inv}, E(h) = t\}.$$

Now let $t \in [0, \ln k]$ be an arbitrary number and let $h \in \Pi_{inv}$ be such that

$$E^1(h, P) = g(t) \quad \text{and} \quad E(h) = t.$$

If $E^1(h) > t$ then we form, for each $s \in [0,1]$, the Bernoulli measure $\mu_s \in \Pi_{inv}$ which satisfies

$$\mu_s(1) = (1-s) \cdot 1 + s \cdot h(1),$$
$$\mu_s(i) = (1-s) \cdot 0 + s \cdot h(i) \qquad \text{for } i = 2, 3, \ldots, k.$$

Now we have

$$- \ln P(1) = E^1(\mu_o, P) \leqslant E^1(\mu_s, P) \leqslant E^1(\mu_1, P) = E^1(h, P) \qquad \forall\, s \in [0,1]$$

and

$$E(\mu_o) = E^1(\mu_o) = 0; \qquad E(\mu_1) = E^1(\mu_1) = E^1(h) > t.$$

Since the function $E^1(\mu_s)$ is continuous with respect to s, there is hence a number $s_o \in [0,1]$ such that

$$E(\mu_{s_o}) = E^1(\mu_{s_o}) = t \qquad \text{and}$$

$$E^1(\mu_{s_o}, P) \leqslant E^1(h, P).$$

This shows that there exists, for each $t \in [0, \ln k]$, a Bernoulli measure μ satisfying

$$g(t) = E^1(\mu, P) = -\sum_{i=1}^{k} \mu(i) \ln P(i) \text{ and } t = E(\mu) = -\sum_{i=1}^{k} \mu(i) \ln \mu(i).$$

In this way the determination of the function $f(t)$, and hence the computation of the Billingsley dimension of the sets X , has been reduced to determining the minimum of the function

$$F(y_1, y_2, \ldots, y_k) := -\sum_{i=1}^{k} y_i \ln P(i), \quad (y_1, y_2, \ldots, y_k) \in \mathbf{R}^k$$

under the constraints

$$y_i \geqslant 0 \qquad \forall\, i = 1, 2, \ldots, k,$$

$$\Sigma_{i=1}^{k}\, y_i = 1,$$

$$\Sigma_{i=1}^{k}\, y_i \ln y_i = -t.$$

The following solution is obtained for the function f(t): On the interval $[0, \ln r]$ the function f(t) increases linearly from 0 to $\frac{\ln r}{-\ln P(1)}$. On the interval $(\ln r, \ln k']$ let

$$P_c(i) := [\Sigma_{j=1}^{k'} P(j)^c]^{-1} P(i) \qquad \forall\, i = 1, \ldots, k' \qquad \forall\, c \geq 0$$

and let c stand for the solution of the equation

$$-\Sigma_{i=1}^{k'} P_c(i) \ln P_c(i) = t, \qquad c \geq 0,$$

(which exists uniquely). Then

$$f(t) = -t \cdot [\Sigma_{i=1}^{k'} P_c(i) \ln P(i)]^{-1} .$$

On the interval$(\ln k', \ln k]$ the function f(t) vanishes identically. In summary, the function f(t) increases monotonically from 0 to 1 on the interval $[0, E(P)]$, being continuously differentiable also at the "point of transition", ln r. On the interval $[E(P), \ln k']$ it decreases monotonically from 1 to $-k' \ln k'[\Sigma_{i=1}^{k'} \ln p(i)]^{-1}$, having a vertical tangent at the point ln k'. Thereafter the function jumps to the value 0.

§ 9. The Billingsley dimension of special sets

In this section we are going to compute the Billingsley dimensions of different types of sets. In Parts 9.A and 9.B these shall be saturated sets characterized by different conditions imposed upon the distribution measures of their points. In Parts 9.C and 9.D we will consider subsets of the smallest saturated sets which can be expressed as intersections of Cantor-type sets.

9.A

Theorem 9.1. Let $H \subset \Pi_{inv}$ be an arbitrary, non-empty set of invariant W-measures. Then the sets

$$S_1(H) := \{x \in X \mid H(x) \subset H\},$$

$$S_2(H) := \{x \in X \mid H \subset H(x)\} \qquad \text{and}$$

$$S_3(H) := \{x \in X \mid H(x) \cap H \neq \emptyset\}$$

satisfy the relations

$$P\text{-dim}(S_1(H)) = P\text{-dim}(S_3(H)) = \sup_{h \in H} \gamma(h, P);$$

$$P\text{-dim}(S_2(H)) = \inf_{h \in H} \gamma(h, P).$$

Proof. We shall write $\cup Y$ for the union of all elements of Y. In this notation we have

$$S_1(H) = \cup \{M_{H'} \mid H' \in \mathbf{H},\ H' \subset H\} \supset \cup \{M_{\{h\}} \mid h \in H\}.$$

By Theorem 8.1 this implies

$$\begin{aligned} P\text{-dim}(S_1(H)) &= \sup \{P\text{-dim}(M_{H'}) \mid H' \in \mathbf{H},\ H' \subset H\} \\ &\leq \sup \{\gamma(h, P) \mid h \in H\} \\ &= P\text{-dim}(\cup \{M_{\{h\}} \mid h \in H\}) \\ &\leq P\text{-dim}(S_1(H)), \end{aligned}$$

from which the assertion on the set $S_1(H)$ follows.

For every point $x \in S_3(H)$ we have

$$P\text{-dim}(M_{H(x)}) = \inf_{h \in H(x)} \gamma(h,P) \leq \inf_{h \in H(x) \cap H} \gamma(h, P) \leq \sup_{h \in H} \gamma(h, P).$$

Therefore, by Theorem 8.1,

$$P\text{-dim}(S_3(H)) \leq \sup_{h \in H} \gamma(h, P).$$

This establishes the assertion on the set $S_3(H)$ since $S_1(H) \subset S_3(H)$.

For every point $x \in S_2(H)$ let

$$P\text{-dim}(M_{H(x)}) = \inf_{h \in H(x)} \gamma(h, P) \leq \inf_{h \in H} \gamma(h, P).$$

On the other hand, the closed convex hull $\overline{\langle H \rangle}$ of H has the following properties:

$H \subset \overline{\langle H \rangle} \in \mathbf{H}$;

$M_{\overline{\langle H \rangle}} \subset S_2(H)$;

$P\text{-dim}(M_{\overline{\langle H \rangle}}) = \inf \{\gamma(h, P) \mid h \in \overline{\langle H \rangle}\} = \inf \{\gamma(h, P) \mid h \in H\}$

(see § 6.E, d) 5.). Combining these two propositions with Theorem 8.1 we obtain

$$P\text{-dim}(S_2(H)) = \inf_{h \in H} \gamma(h, P). //$$

<u>Remark 9.1.1.</u> Since the function $\gamma(h, P)$ is upper semicontinuous, the supremum may be replaced by the maximum in the case of the sets $S_1(H)$ and $S_3(H)$ if the set H is closed. For the set $S_2(H)$ we may always consider the closure $\overline{H}$ instead of H inasmuch as $S_2(\overline{H}) = S_2(H)$.

<u>Remark 9.1.2.</u> If the set H is closed then

$$S_1(H) = \{x \in X \mid \lim_{n \to \infty} d(h_n(x), H) = 0\};$$

$$S_3(H) = \{x \in X \mid \liminf_{n \to \infty} d(h_n(x), H) = 0\}.$$

9.B

We shall now consider saturated subsets of X whose points x are characterized, for a fixed $l \in \mathbb{N}$, by the projections $p^l(h)$ (see § 6.B) of their distribution measures $h \in H(x)$.

For this purpose let the integer $l \in \mathbb{N}$ be fixed and let the metric d^l (see Def. 6.1) be introduced on the set Π^l of W-measures on $(X, \underline{A}^l)$ and $(A^l, \underline{A}^l)$ (see § 6.B), respectively. The following conventions will be useful:

Definition 9.1. For every point $x \in X$ let

$$H^l(x) := p^l(H(x));$$

$$h_n^l(x) := p^l(h_n(x).$$

Furthermore let

$$\mathbf{H}^l := \{H^l \subset \Pi^l_{inv} \mid H^l \neq \emptyset, \text{ closed and connected}\},$$

and for each set $H^l \in \mathbf{H}^l$ let

$$M^l(H^l) := \{x \in X \mid H^l(x) = H^l\}.$$

Remark 9.2.1. For every point $x \in X$ the set $H^l(x)$ belongs to $\mathbf{H}^l$, the projections p^l being continuous.

Remark 9.2.2. The sets $M^l(H^l)$ are saturated.

Remark 9.2.3. $H^l(x)$ is the set of limit points of the sequence $\{h_n^l(x)\}_n$ in the space (Π^l, d^l).

Since the sets $M^l(H^l)$ are saturated, their Billingsley dimensions are given by Theorem 8.1 in principle. A simplified version is given by the following theorem.

Theorem 9.2. If the integer l is larger than the order of the ergodic Markov measure P then, for any $H^l \in \mathbf{H}^l$,

$$\text{P-dim}(M^l(H^l)) = \min_{h \in H^l} \frac{E^l(h)}{E^l(h,P)} .$$

Proof. Let $H^l \in \mathbf{H}^l$ and $c := \inf \{\frac{E^l(h)}{E^l(h,P)} \mid h \in H^l\}$.

For each point $x \in M^l(H^l)$ we have

$$\text{P-dim}(M_{H(x)}) = \inf_{h \in H(x)} \frac{E(h)}{E(h,P)} \leq \inf_{h \in H(x)} \frac{E^l(h)}{E^l(h,P)} = c,$$

using the fact that $E^l(h, P) = E(h, P)$, $E^l(h) \geq E(h)$ and $p^l(H(x)) = H^l$. Therefore it follows by Theorem 8.1 that

$$\text{P-dim}(M^l(H^l)) \leq c.$$

Now let $H := q^l(H^l)$. Then H is the set of those Markov measures of order l - 1 whose restrictions $p^l(h)$ belong to H^l. By the Remarks 6.7.2 and 6.7.3, $H \in \mathbb{H}$ and $p^l(H) = H^l$, thus $M_H \subset M^l(H^l)$. Consequently,

$$P\text{-dim}(M^l(H^l)) \geqslant P\text{-dim}(M_H) = \inf_{h \in H} \frac{E(h)}{E(h,P)} = c,$$

where the last equation uses the Markov property of the W-measures $h \in H$.

Now it only remains to show that the infimum in the definition of c is a minimum. If there exists an $h \in H^l$ with $E^l(h, P) = \infty$ then $c = 0$ and thus the infimum is assumed on H^l.

Otherwise the expression $\frac{E^l(h)}{E^l(h,P)}$, considered as a function of h, is continuous on the compact set H^l by § 6.E, a) 2. and a) 4., which shows that the infimum is assumed on the set H^l in this case also.//

Remark 9.3. From the sets $M^l(H^l)$ further saturated sets may be derived whose Billingsley dimensions may be easily determined by means of Theorem 8.1. Let l be larger than the order of P, let $\emptyset \neq H^l \subset \Pi^l_{inv}$ and

$$S^l_1(H^l) := \{x \in X \mid H^l(x) \subset H^l\},$$

$$S^l_2(H^l) := \{x \in X \mid H^l \subset H^l(x)\},$$

$$S^l_3(H^l) := \{x \in X \mid H^l(x) \cap H^l \neq \emptyset\}.$$

Then it can be shown as in Theorem 9.1 that

$$P\text{-dim}(S^l_1(H^l)) = P\text{-dim}(S^l_3(H^l)) = \sup_{h \in H^l} \frac{E^l(h)}{E^l(h,P)} \quad ;$$

$$P\text{-dim}(S^l_2(H^l)) = \inf_{h \in H^l} \frac{E^l(h)}{E^l(h,P)} \, .$$

If the set H^l is closed then

$$S^l_1(H^l) = \{x \in X \mid \lim_{n\to\infty} d^l(h^l_n(x), H^l) = 0\}.$$

Sets of this form have already been studied and their dimensions determined by Billingsley [9].

Remark 9.4. In order to formulate Theorem 9.2 for real subsets we associate again every number $x \in [0,1]$ with its g-adic digit sequence $\phi(x) = (e_1(x), e_2(x) \ldots)$, where

$g \in \mathbb{N}$, $g \geq 2$, $A = \{0, 1, \ldots, g-1\}$, $X = A^{\mathbb{N}}$ (see Remark 3.3). Now let $l \in \mathbb{N}$, $H^l \in \mathbf{H}^l$ and

$$F^l(H^l) := \{x \in [0,1] \mid H^l(\phi(x)) = H^l\} = \phi^{-1}(M^l(H^l)).$$

Then by Theorem 9.2 and Theorem 3.4,

$$\text{h-dim}(F^l(H^l)) = \min_{h \in H^l} \frac{E^l(h)}{\ln g}.$$

In the case $l = 1$, the set

$$\Pi^1_{inv} = \Pi^1 = \{(y_0, y_1, \ldots, y_{g-1}) \in [0,1]^g \mid \Sigma_{i=0}^{g-1} y_i = 1\}$$

is a convex, closed subset of some hyperplane with $\mathbb{R}^g$. For every point $x \in [0,1]$ and each $n \in \mathbb{N}$, $h^1_n(x)$ is equal to the vector $(h_n(x, 0), h_n(x, 1), \ldots, h_n(x, g-1))$ of relative frequencies of the digits $0, 1, \ldots, g-1$ among the first n digits of the g-adic representation of the number x. Here we have identified the number x with the sequence $\phi(x)$. For every non-empty, closed, connected subset $H^1 \subset \Pi^1$, $F^1(H^1)$ is the set of those numbers $x \in [0,1]$ for which the corresponding sequence $\{(h_n(x, 0), h_n(x, 1), \ldots, h_n(x, g-1))\}_n$ of vectors of relative frequencies possesses the set H^1 as its set of limit points. Then it follows that

$$\text{h-dim}(F^1(H^1)) = \min_{h \in H^1} \frac{E^1(h)}{\ln g}$$

$$= \frac{1}{\ln g} \min \{-\Sigma_{i=0}^{g-1} y_i \mid (y_0, y_1, \ldots, y_{g-1}) \in H^1\},$$

which is equivalent to Satz 2 of Volkmann [49]. The sets which are investigated there in Satz 3 are g-adically saturated and of the form

$$\phi^{-1}(S^1_1(H^1)) = \{x \in [0,1] \mid H^1(x) \subset H^1\}, \; H^1 \subset \Pi^1, \; H^1 \neq \emptyset.$$

According to Theorem 9.2 and Theorem 3.4 they satisfy the equation

$$\text{h-dim}(\phi^{-1}(S^1_1(H^1))) = \frac{1}{\ln g} \sup \{-\Sigma_{i=0}^{g-1} y_i \ln y_i \mid (y_0, y_1, \ldots, y_{g-1}) \in H^1\}$$

in agreement with the result of Volkmann [46].

9.C

In the literature there are many examples (see, e.g., Volkmann [46] and [48], Steinfeld/Wegmann [43]) in which sets are studied whose elements are characterized by the absence of certain digits or blocks of digits in their digit sequences, Cantor's set being the classical example. This set consists of all points of the interval [0,1] which possess a ternary expansion $x = \Sigma_{i=1}^{\infty} 3^{-i} e_i(x)$ in which the digit 1 does not occur, thus having $e_i \in \{0,2\}$ for all i.

Now let a fixed integer $l \in \mathbb{N}$ be chosen and let B be a set of blocks of length l, e.g., $B \subset A^l$. First we generalize the construction of Cantor's set:

Definition 9.2. For every point $x = (x_1, x_2, \ldots) \in A^{\mathbb{N}}$ let

$$\underline{b}_i^l(x) := (x_i, x_{i+1}, \ldots x_{i+l-1}) \in A^l \qquad \forall\, i \in \mathbb{N} \qquad \text{and}$$

$$B^l(x) := \{\underline{b} \in A^l \mid \underline{b} = \underline{b}_i^l(x) \qquad \text{for infinitely many } i \in \mathbb{N}\}.$$

Furthermore let $X_B := \{x \in X \mid \underline{b}_i^l(x) \in B \quad \forall\, i \in \mathbb{N}\}$.

Remark 9.5. If we endow the sequence space $X = A^{\mathbb{N}}$ with the roughest topology in which all cylinders are open, then the set X_B is perfect except for possibly finitely many isolated points, while any set of the form M_H is dense in X.

In this part of § 9 we shall investigate sets of the form $M_H \cap X_B$, i.e. intersections of everywhere dense sets with perfect sets which are, in general, nowhere dense. For this purpose it is appropriate to decompose subsets of A^l into components according to Definition 6.7:

Definition 9.3. An arbitrary, non-empty set $B' \subset A^l$ is called connected if there exists, for any two blocks $\underline{b}, \underline{b}' \in B'$, a finite sequence $x = (x_1, x_2, \ldots, x_{n+l}) \in A^{n+l}$, $n \in \mathbb{N}$, such that

$$\underline{b}_i^l(x) \in B' \qquad \forall\, i = 1, \ldots, n;$$

$$\underline{b}_1^l(x) = \underline{b}; \qquad \underline{b}_n^l(x) = \underline{b}'.$$

The connected components of a subset B' of A^l are the maximal connected subsets of B'.

Remark 9.6.1. If the sets $A_1, \ldots, A_r$ are the components of A^{l-1} relative to B' according to Definition 6.7, then the sets

$$B_j' := \{(b_1, b_2, \ldots, b_l) \in B' \mid (b_1, b_2, \ldots, b_{l-1}) \in A_j, (b_2, b_3, \ldots, b_l) \in A_j\},$$

$j = 1, \ldots, r$, are the connected components of B' in the sense of Definition 9.3.

Remark 9.6.2. For each point $x \in X$, the set $B^l(x) \subset A^l$ is connected since, for sufficiently large i, all blocks $\underline{b}_i^l(x)$ are contained in $B^l(x)$.

Remark 9.6.3. The subset X_B of X is non-empty if and only if the set B contains at least one (non-empty) connected component. For each point $x \in X_B$ the set $B^l(x)$ is completely contained in some connected component B_j of B. In view of

$$\lim_{n\to\infty} h_n(x, \underline{b}) = 0 \qquad \underline{b} \in A^l \setminus B^l(x)$$

one has

$$B^l(h) \subset B^l(x) \subset B_j \qquad \forall\, h \in H(x)$$

(for $B^l(h)$ see Def. 6.5).

The following proposition on the Billingsley dimensions of the sets $M_H \cap X_B$ is valid:

Theorem 9.3. Let $l \in \mathbb{N}$, $B \subset A^l$, $H \in \mathbf{H}$, and assume that the order of P does not exceed $l - 1$. a) If there is a connected component B_j' of $B' := B \cap B^l(P) \subset A^l$ which satisfies

(1) $\quad B^l(h) \subset B_j' \qquad \forall\, h \in H(c),$

then the equation

$$\text{P-dim}(M_H \cap X_B) = \text{P-dim}(M_H \cap X_{B_j'}) = \text{P-dim}(M_H)$$

holds.

b) If none of the connected components B_j' of B' satisfies (1) then

$$\text{P-dim}(M_H \cap X_B) = \text{P-dim}(M_H \cap X_{B_j'}) = 0.$$

Proof. Suppose that case a) holds. Since B_j' is connected, there exists a periodic sequence $x = (x_1, \ldots, x_m, x_1, \ldots, x_m, x_1, \ldots) \in X_{B_j'}$, $m \in \mathbb{N}$, such that $B^l(x) = B_j'$. Then the Markov measure ν of order $l - 1$ satisfying

$$\nu(\underline{b}) = \lim_{n\to\infty} h_n(x, \underline{b}) \qquad \forall\, \underline{b} \in A^l,$$

which turns out to be irreducible and hence ergodic, has the property

$$h <\!*\!< \nu <\!*\!< P \qquad \forall\, h \in H$$

in view of (1) since $B_j' \subset B^l(P)$ and since the ergodic Markov measure P is also of order $l - 1$ (on weak continuity $<\!*\!<$ of two W-measures see Def. 6.4). Hence the assumptions of Theorem 7.2 are satisfied and therefore there exists a W-measure μ on $A^{\mathbb{N}}$ such that

(2) $\quad \mu(M_H) = 1;$

(3) $\quad \frac{\mu}{P}(x) \geqslant \text{P-dim}(M_H) \qquad \forall\, [\mu]\; x \in X;$

(4) $\quad \mu <\!*\!< \nu.$

By (4) one has $\mu(X_{B_j'}) = 1$, thus $\mu(M_H \cap X_{B_j'}) = 1$ by (2), from which it follows by (3) and Theorem 2.12 that

$$\text{P-dim}(M_H \cap X_{B_j'}) \geqslant \text{P-dim}(M_H).$$

Now a monotonicity argument establishes the assertion in case a).

Suppose that case b) occurs. If the intersection of the sets M_H and X_B is empty then the proposition of the theorem is trivial. On the other hand, each point $x \in M_H \cap X_B$ satisfies

$B^1(h) \subset B^1(x) \subset B \qquad \forall\, h \in H$

(see Remark 9.6.3). But since (1) fails to hold for any connected component of $B \cap B^1(P)$, the connected set $B^1(x) \subset A^1$ can not be contained in $B^1(P)$. Hence the point x lies in some P-null-cylinder. This implies

$\text{P-dim}(M_H \cap X_B) = 0$

and hence the assertion for case b) follows by a monotonicity argument.//

Remark 9.7.1. The condition that the order of P should not exceed 1 - 1 does not restrict the applicability of the theorem above inasmuch as the set X_B can always be expressed as $X_{B''}$ with $B'' \subset A^{1''}$, where $1'' \in \mathbb{N}$ is arbitrarily large.

Remark 9.7.2. The theorem just proved is a zero-one law for the Billingsley dimension since either

$\text{P-dim}(M_H \cap X_B) = 0 \cdot \text{P-dim}(M_H)$ or

$\text{P-dim}(M_H \cap X_B) = 1 \cdot \text{P-dim}(M_H)$.

Corollary 9.3.1. (Notation as in Theorem 9.3). The set $M_H \cap X_B$ is non-empty if and only if there exists a connected component B_j of B such that

(1') $\quad B^1(h) \subset B_j \qquad \forall\, h \in H.$

Proof. If $x \in M_H \cap X_B$ then $H(x) = H$ and hence there exists by Remark 9.6.3 a connected component B_j of B which satisfies (1'). On the other hand, if (1') is valid then all assumptions of part a) of Theorem 9.3 are satisfied for $B_j' = B_j$ with P being replaced by the measure P_g of equidistribution since $B^1(P_g) = A^1$. The proof of this part involves a W-measure μ with $\mu(M_H \cap X_{B_j'}) = 1$. Thus, in particular, $M_H \cap X_B \neq \emptyset$. //

As a further corollary of Theorem 9.3 it is possible to extend the relation (SUP) of Theorem 8.1.b) to a family of subsets of X which contains saturated sets as a proper sub-family.

Corollary 9.3.2. a) For each $\alpha \in I$ of an arbitrary set I of indices let $H_\alpha \in \mathbf{H}$, $1_\alpha \in \mathbb{N}$ and $B_\alpha \in A^{1_\alpha}$. Then

$$\text{P-dim}\Big(\bigcup_{\alpha\in I} M_{H_\alpha} \cap X_{B_\alpha}\Big) = \sup_{\alpha\in I} \text{P-dim}(M_{H_\alpha} \cap X_{B_\alpha}).$$

b) For each $\alpha \in I$ of an arbitrary set I of indices let Y_α be an arbitrary union of sets of the form $M_H \cap X_B$ with $H \in \mathbf{H}$, $1 \in \mathbb{N}$ and $B \subset A^1$. Then the following relation holds:

(SUP) $\quad \text{P-dim}\Big(\bigcup_{\alpha\in I} Y_\alpha\Big) = \sup_{\alpha\in I} \text{P-dim}(Y_\alpha).$

Proof. It suffices to establish part a) since part b) is an immediate consequence.

In order to establish proposition a) let

$$I' := \{\alpha \in I \mid P\text{-dim}(M_{H_\alpha} \cap X_{B_\alpha}) = P\text{-dim}(M_{H_\alpha})\}.$$

It follows from the proof of Theorem 9.3.b) that each point of the set

$\bigcup_{\alpha \in I \setminus I'} M_{H_\alpha} \cap X_{B_\alpha}$ is contained in some P-null-cylinder. Therefore

$$P\text{-dim}(\bigcup_{\alpha \in I} M_{H_\alpha} \cap X_{B_\alpha}) = P\text{-dim}(\bigcup_{\alpha \in I'} M_{H_\alpha} \in X_{B_\alpha}) \leqslant P\text{-dim}(\bigcup_{\alpha \in I'} M_{H_\alpha})$$

$$= \sup_{\alpha \in I'} P\text{-dim}(M_{H_\alpha}) = \sup_{\alpha \in I'} P\text{-dim}(M_{H_\alpha} \cap X_{B_\alpha})$$

$$= \sup_{\alpha \in I} P\text{-dim}(M_{H_\alpha} \cap X_{B_\alpha}) \leqslant P\text{-dim}(\bigcup_{\alpha \in I} M_{H_\alpha} \cap X_{B_\alpha}).$$

Therefore equality must hold everywhere in this sequence of inequalities, which proves part a).//

By means of Theorem 9.3 it is also easy to determine the Billingsley dimension of the set X_B itself. Before doing so we shall join into one set those invariant W-measures which shall be essential in this context:

Definition 9.4. Let $l \in \mathbb{N}$ and $B \subset A^l$. Then

$$\Pi(B) := \{\mu \in \Pi_{inv} \mid B^l(\mu) \subset B\}.$$

Remark 9.8.1. The set $\Pi(B)$ is a convex, closed subset of Π_{inv}.

Remark 9.8.2. For each invariant W-measure $\mu \in \Pi_{inv}$ the following relation holds:

$$\mu \in \Pi(B) \Leftrightarrow \mu(X_B) = 1.$$

Remark 9.8.3. For each point $x \in X_B$ we have $H(x) \subset \Pi(B)$.

Remark 9.8.4. The set $\Pi(B)$ is non-empty if and only if $X_B \neq \emptyset$.

Remark 9.8.5. If ν is a Markov measure of order $l - 1$ then

$$\Pi(B^l(\nu)) = \{\mu \in \Pi_{inv} \mid \mu <\!*\!< \nu\}.$$

(For weak continuity <*< of two W-measures see Def. 6.4; compare Remark 6.6.2).

Corollary 9.3.3. Let $l \in \mathbb{N}$, $B \subset A^l$ and suppose the order of P does not exceed $l - 1$. Then all of the following quantities are equal:

$c_1 := P\text{-dim}(X_B)$;

$c_2 := \max \{\gamma(h, P) \mid h \in \Pi(B)\}$;

$c_3 := \max \{\gamma(\nu, P) \mid \nu \in \Pi(B),\ \nu \text{ ergodic, Markov, order } l - 1\}$.

Proof. By reason of compactness and upper semicontinuity the maxima c_2 and c_3 are assumed. In the case $X_B = \emptyset$ we have trivially $c_1 = c_2 = c_3 = 0$.

Otherwise we have with $H(x) \subset \Pi(B)$ for each point $x \in X_B$. And therefore, by Theorem 8.1,

$$c_1 = P\text{-dim}(X_B) \leq \sup \{P\text{-dim}(M_H) \mid H \in \mathbb{H}, H \subset \Pi(B)\} = c_2.$$

But for each invariant W-measure $h \in \Pi(B)$, the Markov measure ν of order $l - 1$ with

$$\nu(\underline{b}) = h(\underline{b}) \qquad \forall\, \underline{b} \in A^l$$

and its ergodic components $\nu_1, \nu_2, \ldots, \nu_r$ (see § 6.C) are also contained in $\Pi(B)$. If the measure ν possesses the representation $\nu = \Sigma_{i=1}^r \alpha_i \nu_i$ with $\alpha_i > 0$, $\Sigma_{i=1}^r \alpha_i = 1$, then

$$E(h) \leq E(\nu) = \Sigma_{i=1}^r \alpha_i E(\nu_i) \qquad \text{and}$$

$$E(h, P) = E(\nu, P) = \Sigma_{i=1}^r \alpha_i E(\nu_i, P), \qquad \text{hence}$$

$$\gamma(h, P) \leq \gamma(\nu, P) \leq \max \{\gamma(\nu_i, P) \mid i = 1, 2, \ldots, r\} \leq c_3.$$

This implies $c_2 \leq c_3$.

Finally let $\nu \in \Pi(B)$ be an ergodic Markov measure of order $l - 1$. Then $\nu(X_B) = 1$ by Remark 9.8.2; furthermore, $\gamma(\nu, P) \leq P\text{-dim}(X_B)$ by Example 2.4. Hence $c_3 \leq c_1$, which proves the assertion.//

Remark 9.9. The equation

$$P\text{-dim}(X_B) = \max \{\gamma(\nu, P) \mid \nu \in \Pi(B), \nu \text{ ergodic}\}$$

holds, where the number on the right-hand side is clearly between the constants c_2 and c_3 of Corollary 9.3.3. Therefore, in the case of the measure P_g of equidistribution on $A^{\mathbb{N}}$, the quantity $\ln |A| \cdot P_g\text{-dim}(X_B)$ is equal to the maximal entropy which an ergodic (invariant) W-measure on X_B may possess.

In the following example we shall investigate further the Billingsley dimension of the set X_B for which various representations have been given in Corollary 9.3.3.

Example 9.1. Let P be an irreducible Markov measure of order 1 and let $B \subset A^2$ be a set of blocks of length 2. Then X_B is, by Definition 9.2., the set of points $x = (x_1, x_2, \ldots) \in A^{\mathbb{N}}$ for which all blocks of the form (x_i, x_{i+1}), $i \in \mathbb{N}$, belong to B.

1. By Corollary 9.3.3 we have

(1) $\quad P\text{-dim}(X_B) = \max \gamma(\nu, P)$,

where the maximum is extended over all ergodic Markov measures ν of order 1 whose "support" $B^2(\nu) = \{\underline{b} \in A^2 \mid \nu(\underline{b}) > 0\}$ is a subset of B. Trivially we have $\gamma(\nu, P) = 0$

for those Markov measures among them which fail to be weakly continuous with respect to P, i.e., which violate one of the conditions "$\nu <\!*\!< P$" or "$B^2(\nu) \subset B^2(P)$" (which are equivalent according to Remark 6.6.2). For the remaining Markov measures we have

$$B^2(\nu) \subset B \cap B^2(P).$$

By Remark 6.4 the set $B^2(\nu)$ is connected since ν is ergodic. Therefore, $B^2(\nu)$ is completely contained in one of the connected components of $B \cap B^2(P)$ (see Def. 9.3.), and thus we have established the following proposition:

If $B \cap B^2(P)$ does not have connected components then $P\text{-dim}(X_B) = 0$. Otherwise it suffices to extend the maximum in Equation (1) over all ergodic Markov measures of order 1 for which $B^2(\nu)$ is completely contained in some connected component of $B \cap B^2(P)$.

2. In the sequel we shall assume that $B \cap B^2(P)$ does possess connected components. Let ν be an ergodic Markov measure of order 1 for which the maximum in (1) is attained and let B^* be the connected component of $B \cap B^2(P)$ to which $B^2(\nu)$ belongs. Then we have

$$(2) \qquad B^2(\nu) = B^*.$$

Indeed, suppose (2) would not be satisfied. Then we choose a Markov measure μ of order 1 for which $B^2(\mu) = B$. (It was shown in the proof of Theorem 9.3 how such a Markov measure can always be constructed). For each $\alpha \in [0,1]$ let μ_α be the Markov measure of order 1 for which

$$\mu_\alpha(\underline{b}) = \alpha\mu(\underline{b}) + (1-\alpha)\,\nu(\underline{b}) \quad \forall\ \underline{b} \in A^2.$$

First we consider the numerator in the fraction

$$(3) \qquad \gamma(\mu_\alpha, P) = \frac{E(\mu_\alpha)}{E(\mu_\alpha, P)} .$$

Clearly we have (compare § 6.E)

$$E(\mu_\alpha) = E^2(\mu_\alpha) = -\sum_{i\in A}\ \sum_{j\in A}\ \mu_\alpha(i,j)\ \ln \frac{\mu_\alpha(i,j)}{\mu_\alpha(i)} .$$

Since we have assumed that (2) should not be valid, there exists a block $(i,j) \in A^2$ such that

$$\mu_\alpha(i) > 0 \qquad \forall\ \alpha \in [0,1],$$
$$\mu_\alpha(i,j) = \alpha\ \mu(i,j) \qquad \forall\ \alpha \in [0,1],$$
$$\mu(i,j) > 0 .$$

For the derivative of the numerator in the expression (3) this implies

$$\lim_{\alpha\to 0^+} \frac{d}{d\alpha} E(\mu_\alpha) = +\infty.$$

On the other hand, since the derivative of the denominator in the expression (3) with respect to α is constant, we also have

$$\lim_{\alpha\to 0^+} \frac{d}{d\alpha} \gamma(\mu_\alpha, P) = +\infty.$$

Thus there exists a number $\alpha \in (0,1)$ such that

$$\gamma(\mu_\alpha, P) > \gamma(\nu, P).$$

Inasmuch as these measures μ_α are also ergodic, this contradicts the maximum property of ν which we have assumed. This establishes (2).

3. Now let μ be an arbitrary Markov measure of order 1 with $B^2(\mu) \subset B^*$. Then the quantities

$$x_{ij} := \mu(i, j) \qquad \forall\ (i, j) \in A^2$$

satisfy the relations

$$(4) \quad \begin{cases} x_{ij} \geqslant 0 & \forall\ (i, j) \quad A, \\ x_{ij} = 0 & \forall\ (i, j) \in A^2 \setminus B^*. \end{cases}$$

$$(5) \qquad \sum_{j\in A} x_{ij} = \sum_{j\in A} x_{ji} \qquad \forall\ i \in A,$$

$$(6) \qquad \sum_{i\in A} \sum_{j\in A} x_{ij} = 1.$$

Furthermore, we shall use the following abbreviations:

$$p_i := P(i) \qquad \forall\ i \in A$$

$$p_{ij} := P(i, j) \qquad \forall\ (i, j) \in A$$

$$Z := Z(x_{ij}) = \sum_{i\in A} \sum_{j\in A} x_{ij} (\ln x_{ij} - \ln \sum_{\ell\in A} x_{i\ell})$$

$$N := N(x_{ij}) = \sum_{i\in A} \sum_{j\in A} x_{ij} (\ln p_{ij} - \ln p_i)$$

$$f := f(x_{ij}) = \frac{Z}{N}.$$

With this notation the right-hand side of the Equation (1) is equal to the maximum of the function f under the constraints (4), (5) and (6). This maximum is attained at the point

$$x_{ij}^0 := \nu(i, j) \qquad \forall\ (i, j) \in A^2.$$

Because of (2) this point is contained in the interior of the domain which is defined by the given constraints. By Lagrange's method of multipliers there exist real parameters λ and λ_i, $i \in A$, such that the auxiliary function

$$\phi((x_{ij})_{(i,j)\in B^*}) := f + \sum_i \lambda_i(\sum_j^* x_{ij} - \sum_j^* x_{ji}) + \lambda(\sum_{i,j}^* x_{ij} - 1)$$

satisfies the equation

$$\frac{\partial}{\partial x_{rs}} \phi((x^o_{ij})_{(i,j)\in B^*}) = 0 \qquad \forall\ (r,s) \in B^*.$$

Here Condition (4) is observed by restricting all summations to those $i \in A$ and $j \in A$ for which the block (i, j) belongs to B^*. The same modification is carried out in (5), (6) and in the definition of Z and N. We express this restriction in the range of summation by an asterisk, regardless of the notation used for the initial indices.

The partial derivatives with respect to x_{rs} with $(r, s) \in B^*$ of the expressions involved in the function ϕ are as follows:

$$\frac{\partial}{\partial x_{rs}} Z = \ln x_{rs} - \ln \sum_{\ell}^* x_{r\ell}\ ,$$

$$\frac{\partial}{\partial x_{rs}} N = \ln p_{rs} - \ln p_r\ ,$$

$$\frac{\partial}{\partial x_{rs}} \sum_i \lambda_i(\sum_j^* x_{ij} - \sum_j x_{ji}) = \lambda_r - \lambda_s\ ,$$

$$\frac{\partial}{\partial x_{rs}} \lambda(\sum_{i,j}^* x_{ij} - 1) = \lambda.$$

Thus we obtain

$$(7)\quad \left\{ \begin{aligned} &\frac{\partial}{\partial x_{rs}} \phi((x^o_{ij})_{(i,j)\in B^*}) \\ &= \frac{1}{N^2} \{(\ln x^o_{rs} - \ln \sum_{\ell}^* x^o_{r\ell})\, N - (\ln p_{rs} - \ln p_r)\, Z\} \\ &\quad + \lambda_r - \lambda_s + \lambda \\ &= 0 \qquad\qquad \forall\ (r, s) \in B^*. \end{aligned} \right.$$

In order to verify that $\lambda = 0$ we multiply equation (7) by x^o_{rs}, summing the results over all $(r, s) \in B^*$. Then the expressions in braces add up to $ZN - NZ = 0$ such that the equation

$$\sum_{r,s}^{*} \lambda_r x^o_{rs} - \sum_{r,s}^{*} \lambda_s x^o_{rs} + \sum_{r,s}^{*} \lambda x^o_{rs} = 0$$

remains. In view of the constraint (5) the first two summations cancel each other, and finally the constraint (6) implies

$$\lambda = 0.$$

This simplifies the equation (7) which is then solved by

$$(8) \qquad \frac{x^o_{rs}}{\sum_{\ell}^{*} x^o_{r\ell}} = \left(\frac{p_{rs}}{p_r}\right)^c \cdot \frac{e^{\lambda_s}}{e^{\lambda_r}} \qquad \forall\ (r, s) \in B^*,$$

where

$$c := \frac{Z}{N} = f((x^o_{ij})_{i,j\in B^*}) .$$

Summing over all $s \in A$ with $(r, s) \in B^*$ yields the relation

$$e^{\lambda_r} = \sum_{s}^{*} \left(\frac{p_{rs}}{p_r}\right)^c e^{\lambda_s}.$$

This equation may be interpreted as follows:

The matrix $(p^*_{ij}(c))_{(i,j)\in A^2}$, where

$$(9) \qquad p^*_{ij}(c) = \begin{cases} \left(\frac{p_{ij}}{p_i}\right)^c & \text{for } (i,j) \in B^* \\ 0 & \text{for } (i,j) \in A^2 \setminus B^* \end{cases}$$

possesses the eigenvalue 1, and there is a non-negative column-vector in the corresponding space of eigenvectors.

4. Now let B^* be an arbitrary connected component of $B \cap B^2(P)$, let $c \in \mathbb{R}$ and suppose the matrix $(p^*_{ij}(c))_{(i,j)\in A^2}$, defined as in (9), has a non-negative column-eigenvector $(c_j)_{j\in A}$ associated with the eigenvalue 1. Then we have

$$(10) \qquad c_i = \sum_{j\in A} p^*_{ij}(c)\ c_j \qquad \forall\ i \in A$$

and thus

$$\sum_i c_i = \sum_j \left(\sum_i p^*_{ij}(c)\right) c_j.$$

Now if $c_j > 0$ then there exists an index $i \in A$ such that $p^*_{ij}(c) > 0$. On the other hand we would have $p^*_{ij}(c) > 1$ for $c < 0$. Therefore $c \geqslant 0$. If we further observe that $c_i > 0$ and $c_j > 0$ for all $(i, j) \in B^*$ then it can be seen from (10) that a matrix of transition probabilities is defined by

$$y_{ij} := \begin{cases} p^*_{ij}(c)\,\frac{c_j}{c_i} & \text{for } (i, j) \in B^*, \\ 0 & \text{for } (i, j) \in A^2 \setminus B^*. \end{cases}$$

For this matrix there exists a uniquely determined Markov measure μ of order 1 (compare, e.g., Billingsley [11, Example 3.1]) such that the quantities

$$x_{ij} := \mu(i, j) \qquad \forall\, (i, j) \in A^2$$

satisfy the relations

$$y_{ij} = \frac{x_{ij}}{\sum\limits_{\ell\in A} x_{i\ell}} .$$

Thus the numbers x_{ij} satisfy the equation (8) and furthermore the equations (4), (5) and (6). Hence, substituting the point $(x_{ij})_{(i,j)\in A^2}$ in the definition of f, we obtain $\gamma(\mu, P) = c$. On the one hand this implies

$$P\text{-dim}(X_B) \geqslant c$$

and on the other hand,

$$c \leqslant 1.$$

5. The following proposition is obtained immediately from steps 3 and 4 :
Let B^{**} be the union of the connected components of $B \cap B^2(P)$ and let

$$p^{**}_{ij}(c) := \begin{cases} \left(\frac{P(i,j)}{P(i)}\right)^c & \text{for } (i, j) \in B^{**} \\ 0 & \text{for } (i, j) \in A^2 \setminus B \end{cases},$$

and let C be the set of all real numbers c for which the matrix $(p^{**}_{ij}(c))_{(i,j)\in A^2}$ possesses a non-negative column-vector as eigenvector associated with the eigenvalue 1. If the set B^{**} is empty then $P\text{-dim}(X_B) = 0$. If B^{**} is non-empty then the same is true for the set C. Furthermore, C is bounded from above and we have

$$P\text{-dim}(X_B) = \max C.$$

6. The Perron-Frobenius Theorem (see Frobenius [25]) implies that the matrix $(p^{**}_{ij}(c))$, being non-negative and decomposed into connected components, possesses at least one positive eigenvalue for every number $c \in \mathbb{R}$. The largest positive eigenvalue λ_c possesses a non-negative eigenvector and exceeds all other eigenvalues in modulus. On the other hand, λ_c is the maximum of all $\lambda > 0$ for which a non-negative vector $(c_j)_{j\in A} \neq (0)_{j\in A}$ exists such that

$$\lambda c_i \leqslant \sum_{j\in A} p^{**}_{ij}(c)\, c_j \qquad \forall\, i \in A.$$

From this representation it can be seen that the function λ_c decreases monotonically as c increases. Therefore the Billingsley dimension of the set X_B may also be described as follows: The number P-dim(X_B) equals the maximum of all real numbers c for which the matrix $(p_{ij}^{**}(c))_{(i,j)\in A^2}$ possesses the eigenvalue 1, where the maximum of the empty set has to be defined as 0.

7. Now let $\ell \in \mathbb{N}$, $\ell \geqslant 2$, and let P be an irreducible Markov measure of order $\ell - 1$. We consider a set $B \subset A^\ell$ of blocks of length ℓ and the corresponding set X_B of points $x = (x_1, x_2, \ldots) \in X$ for which all blocks $(x_{i+1}, \ldots, x_{i+\ell})$, $i \in \mathbb{N}_0$, belong to B. In order to use the arguments of the preceding steps we have to replace the set A by the set $\hat{A} := A^{\ell-1}$, the indices $i, j \in A$ by $\underline{i}, \underline{j} \in \hat{A}$, the set B^{**} by

$$\hat{B}^{**} = \{(\underline{i}, \underline{j}) \in \hat{A}^2 \mid \underline{i} = (i_1, \ldots, i_{\ell-1}), \underline{j} = (i_2, \ldots, i_{\ell-1}) \text{ where}$$
$$(i_1, \ldots, i_\ell) \text{ belongs to a connected component of } B \cap B^\ell(P)\};$$

finally the matrix $(p_{ij}^{**}(c)_{(i,j)\in A^2}$ has to be replaced by the matrix $(\hat{p}_{\underline{ij}}^{**}(c)_{(\underline{i},\underline{j})\in A^2}$, where

$$\hat{p}_{\underline{ij}}^{**}(c) = \begin{cases} \left(\frac{P(i_1,\ldots,i_\ell)}{P(i_1,\ldots,i_{\ell-1})}\right)^c, & \text{if } (\underline{i}, \underline{j}) \in \hat{B}^{**} \\ 0 & \text{otherwise} \end{cases}$$

with $\underline{i}$ and $\underline{j}$ defined as above. Thus we obtain the analagous proposition as in step 6: We have P-dim(X_B) equal to the maximum of those real numbers c for which the matrix $(\hat{p}_{\underline{ij}}^{**}(c))_{(\underline{i},\underline{j}) \in \hat{A}^2}$ possesses the eigenvalue 1. (Here again we have to let max $\emptyset = 0$).

Remark 9.10. The preceding example generalizes Satz 1 of Volkmann [46] and Satz 4 of Volkmann[48] with respect to the concept of dimension used. The latter result has also been generalized by Steinfeld/Wegmann [43] in the sense of Billingsley dimensions with respect to a Markov measure. As the proofs of Volkmann and Steinfeld/Wegmann, our Example 9.1 also involves a theorem of Frobenius [25] on the eigenvalues of non-negative matrices. But this theorem is not needed to prove the representation of the Billingsley dimension given in step 5 of Example 9.1.

9.D

Generalizing the investigation of § 9.C let, for every $l \in \mathbb{N}$, a set $B_l \subset A^l$ be given and define

$$Y := \bigcap_{l \in \mathbb{N}} X_{B_l}.$$

Without changing the set Y, each set B_l may be replaced by a subset (to be called B_l again) in such a way that the sequences $\{X_{B_l}\}_l$ and $\{\Pi(B_l)\}_l$ are monotonically decreasing. With this modification, letting

$$\Pi_o := \bigcap_{l \in \mathbb{N}} \Pi(B_l),$$

the following theorem holds:

Theorem 9.4. The following quantities are equal to each other:

$$c_1 := \text{P-dim}(Y);$$

$$c_2 := \lim_{l \to \infty} \text{P-dim}(X_{B_l});$$

$$c_3 := \max \{\gamma(h, P) \mid h \in \Pi_o\};$$

$$c_4 := \max \{\gamma(\nu, P) \mid \nu \in \Pi_o, \nu \text{ ergodic}\}.$$

Proof. If the set $\Pi(B_l)$ is empty for some $l \in \mathbb{N}$ then the same is true for the sets X_{B_l}, Y and Π_o. Thus the assertion is trivially true for $c_1 = c_2 = c_3 = c_4 = 0$.

Otherwise $\Pi(B_l) \neq \emptyset$ for all $l \in \mathbb{N}$. Then the set Π_o, being the intersection of a monotonically decreasing sequence of non-empty,compact sets, is itself non-empty and compact. First we have $c_1 \leq c_2$ on account of monotonicity. For each $l \in \mathbb{N}$ let $h_l \in \Pi(B_l)$ such that $\gamma(h_l, P) = \text{P-dim}(X_{B_l})$. The existence of such W-measures h_l is guaranteed by Corollary 9.3.3. Since the space under consideration is compact, the sequence $\{h_l\}_l$ has at least one limit point h_o which must belong to Π_o. Thus it follows from the upper semicontinuity of the function $\gamma(\cdot, P)$ that

$$c_2 = \lim_{l \to \infty} \text{P-dim}\ \gamma(h_l, P) \leq \gamma(h_o, P) \leq c_3.$$

The maximum in the definition of c_3 is assumed since Π_o is compact.

Now let $\mu \in \Pi_o$ be an invariant W-measure and let μ be the W-distribution (see § 1.B) induced by μ on Π_{inv}. Then the following propositions follow from Lemma 1.2:

(1) $\tilde{\mu}$-almost all W-measures $\nu \in \Pi_{inv}$ are ergodic;

(2) $\mu(\underline{b}) = \int \nu(\underline{b})\, d\tilde{\mu} \quad \forall\, \underline{b} \in A^l \quad \forall\, l \in \mathbb{N}$;

(3) $E(\mu) = \int E(\nu)\, d\tilde{\mu}$.

The representation (2) now implies

(4) $\quad \nu \in \Pi_0 \qquad\qquad\qquad \forall\, [\tilde{\mu}]\ \nu \in \Pi_{inv}$;

(5) $\quad E(\mu, P) = \int E(\nu, P)\, d\tilde{\mu}$.

Furthermore it follows from (3) and (5) that

$$\tilde{\mu}(\{\nu \in \Pi_{inv} \mid \gamma(\mu, P) \leqslant \gamma(\nu, P)\}) > 0.$$

Therefore there exists on account of (1) and (4) an ergodic W-measure $\nu \in \Pi_0$ such that $\gamma(\mu, P) \leqslant \gamma(\nu, P)$. This shows that $c_3 \leqslant c_4$.

For each ergodic W-measure $\nu \in \Pi_0$ we have $\nu(X_{B_l}) = 1$ for all $l \in \mathbb{N}$ (compare Remark 9.8.2), hence $\nu(Y) = 1$. In analogy to Example 2.4 this implies $\gamma(\nu, P) \leqslant c_1$ and finally $c_4 \leqslant c_1$. This completes the proof of the theorem.//

Remark 9.11. Since the assertion $h \in \Pi_0$ and $h(Y) = 1$ are equivalent in this case too, one has

$$P\text{-dim}(Y) = \max\,\{\gamma(\nu, P) \mid \nu \in \Pi_{erg},\ \nu(Y) = 1\}.$$

In the case of the measure P_g of equidistribution the quantity $\ln |A| \cdot P_g\text{-dim}(Y)$ is again equal to the maximal entropy which an ergodic (invariant) W-measure defined on Y may assume.

Final remark. By means of the theorems and methods developed in Parts A through D of this section the computation of the Billingsley dimension (and Hausdorff dimension, in particular) of a given set is frequently reduced to a multidimensional extremal value problem with constraints which in turn may be solved by methods of calculus. For example, Satz 1 of Volkmann [45] and the Digit Theorem of Volkmann [47], which generalizes this theorem as well as results of Eggleston [23] and [24], may be established essentially by means of § 9.B. Hausdorff dimensions may be replaced here by more general Billingsley dimensions. It was explained in Remark 9.4 how Satz 1 of Volkmann [49] may be derived.

It was shown by Example 9.1 how § 9.C, among other applications, allows a generalization of Satz 1 of Volkmann [46] and Satz 4 of Volkmann [48] (compare Remark 9.10). The case of infinitely many conditions in terms of blocks of digits, however, as investigated in § 9.D amounts to a generalization of Satz 4 of Volkmann [48] in an additional direction.

REFERENCES

[1] A. Baker, W.M. Schmidt: Diophantine approximation and Hausdorff dimension. Proc. London Math. Soc. 21 (1970), 1-11.

[2] R.C. Baker: Dyadic Methods in the measure theory of numbers. Transactions of the AMS 221 (1976), 419-432.

[3] H. Bauer: Wahrscheinlichkeitstheorie und Grundzüge der Maßtheorie. De Gruyter, Berlin, 1968.

[4] A.F. Beardon: On the Hausdorff dimension of general Cantor sets. Proc. Camb. Phil. Soc. 61 (1965), 679-694.

[5] C. Berge: Graphs and Hypergraphs. North-Holland Publishing Company, Amsterdam, 1973.

[6] M. Bernay: La dimension de Hausdorff de l'ensemble des nombres r-déterministes. C.R. Acad. Sc. Paris, Série A 280 (1975), 539-541.

[7] A.S. Besicovitch: On the sum of digits of real numbers represented in the dyadic system. Math. Ann. 110 (1934), 321-330.

[8] W.A. Beyer: Hausdorff dimension of level sets of some Rademacher series. Pacific J. 12 (1962), 35-46.

[9] P. Billingsley: Hausdorff dimension in probability theory. Ill. J. Math. 4 (1960), 187-209.

[10] P. Billingsley: Hausdorff dimension in probability theory II. Ill. J. Math. 5 (1961), 291-298.

[11] P. Billingsley: Ergodic Theory and Information. Wiley, New York, 1965.

[12] P. Billingsley: Convergence of Probability Measures. Wiley, New York, 1968.

[13] P. Billingsley, I. Henningsen: Hausdorff dimension of some continued-fraction sets. Z. Wahrscheinlichkeitstheorie verw. Geb. 31 (1975), 163-173.

[14] D.W. Boyd: The residual set dimension of the Apollonian packing. Mathenatika 20 (1973), 170-174.

[15] R.J. Buck: A generalized Hausdorff dimension for functions and sets. Pacific J. 33 (1970), 69-78.

[16] R.J. Buck: Hausdorff dimensions for compact sets in $\mathbb{R}^n$. Pacific J. 44 (1973), 421-434.

[17] G. Choquet: Lectures on Analysis, Vol. II. Benjamin, New York, 1969.

[18] J. Cigler: Ziffernverteilung in ϑ-adischen Brüchen. Math. Zeitschr. 75 (1961), 8-13.

[19] J. Cigler: Hausdorffsche Dimensionen spezieller Punktmengen. Math. Zeitschr. 76 (1961), 22-30.

[20] C.M. Colebrook: The Hausdorff dimension of certain sets of nonnormal numbers. Michigan Math. J. 17 (1970), 103-116.

[21] M. Denker, C. Grillenberger, K. Sigmund: Ergodic Theory on Compact Spaces. Springer, Berlin, 1976.

[22] J. Dugundji: Topology. Allyn and Bacon, Boston, 1974.

[23] H.G. Eggleston: The fractional dimension of a set defined by decimal properties. J. Math. 20 (1949), 31-36.

[24] H.G. Eggleston: Sets of fractional dimension which occur in some problems of number theory. Proc. London Math. Soc. 254 (1951), 42-93.

[25] G. Frobenius. Über Matrizen aus positiven Elementen. Sitzungsber. preuß. Akad. Wiss., Physik.-math. Kl. 26 (1908), 471-476.

[26] J. Galambos: Representations of Real Numbers by Infinite Series. Springer, Berlin, 1976.

[27] H. Gericke: Theorie der Verbände. Bibliographisches Institut, Mannheim, 1967.

[28] I.J. Good: The Estimation of Probabilities. M.I.T. Press, Cambridge, 1965.

[29] K. Hatano: Evaluation of Hausdorff measures of generalized Cantor sets. J. Sci. Hiroshima Univ. Ser. A-I 32 (1968), 371-379.

[30] J. Hawkes: On the Hausdorff dimension of the intersection of the range of a stable process with a Borel set. Z. Wahrscheinlichkeitstheorie verw. Geb. 19 (1971), 90-102.

[31] J.R. Kinney, T.S. Pitcher: The dimension of some sets defined in terms of f-expansions. Z. Wahrscheinlichkeitstheorie verw. Geb. 4 (1966), 293-315.

[32] V. Knichal: Dyadische Entwicklungen und Hausdorffsches Maß. Mêm. Soc. Roy. Sci. Bohême, Cl. Sci. Nr. XIV (1933).

[33] K. Kuratowski: Topology. Academic Press, New York, 1966/1968.

[34] K. Nagasaka: La dimension de Hausdorff de certains ensembles dans [0,1]. Proc. Japan Acad. Ser. A 54 (1978), 109-112.

[35] K.R. Parthasarathy: Probability Measures on Metric Spaces. Academic Press, New York, 1967.

[36] J. Peyrière: Calculs de dimensions de Hausdorff. Duke Math. J. 44 (1977), 591-601.

[37] A.D. Pollington: The Hausdorff dimension of certain sets related to sequences which are not dense mod 1. Quart. J. Math. Oxford 31 (1980), 351-361.

[38] G. Rauzy: Nombres normaux et processus déterministes. Acta Arithmetica 29 (1976), 211-225.

[39] C.A. Rogers: Hausdorff Measures. Cambridge Univ. Press 1970.

[40] F. Schweiger: Abschätzung der Hausdorffdimension für Mengen mit vorgeschriebenen Häufigkeiten der Ziffern. Monatshefte Math. 76 (1972), 138-142.

[41] F. Schweiger, W. Stradner: Die Billingsleydimension von Mengen mit vorgeschriebenen Ziffern. Sb. Öst. Ak. Wiss. Math.-naturw. Kl., Abt. II, 180 (1971), 95-109.

[42] M. Sion, R.C. Willmott: Hausdorff measures on abstract spaces. Trans. AMS 123 (1966), 275-309.

[43] L. Steinfeld, H. Wegmann: Die Dimension von Teilmengen eines Wahrscheinlichkeitsraumes. Math. Ann. 184 (1970), 317-325.

[44] B. Volkmann: Über Klassen von Mengen natürlicher Zahlen. J. reine angew. Math. 190 (1952), 199-230.

[45] B. Volkmann: Über Hausdorffsche Dimensionen von Mengen, die durch Ziffereigenschaften charakterisiert sind. II. Math. Zeitschr. 59 (1953), 247-254.

[46] B. Volkmann: Über Hausdorffsche Dimensionen von Mengen, die durch Zifferneigenschaften charakterisiert sind. III. Math. Zeitschr. 59 (1953), 259-270.

[47] B. Volkmann: Über Hausdorffsche Dimensionen von Mengen, die durch Zifferneigenschaften charakterisiert sind. IV. Math. Zeitschr. 59 (1954), 425-433.

[48] B. Volkmann: Über Hausdorffsche Dimensionen von Mengen, die durch Zifferneigenschaft charakterisiert sind. V. Math. Zeitschr. 65 (1956), 389-413.

[49] B. Volkmann: Über Hausdorffsche Dimensionen von Mengen, die durch Zifferneigenschaften charakterisiert sind. VI. Math. Zeitschr. 68 (1958), 439-449.

[50] H. Wegmann: Über den Dimensionsbegriff in Wahrscheinlichkeitsräumen. Z. Wahrscheinlichkeitstheorie verw. Geb. 9 (1968), 216-221.

[51] H. Wegmann: Über den Dimensionsbegriff in Wahrscheinlichkeitsräumen. II. Z. Wahrscheinlichkeitstheorie verw. Geb. 9 (1968), 222-231.

INDEX OF SYMBOLS

The numbers refer to the pages where the symbols are introduced.

INDEX OF TERMS

Vol. 728: Non-Commutative Harmonic Analysis. Proceedings, 1978. Edited by J. Carmona and M. Vergne. V, 244 pages. 1979.

Vol. 729: Ergodic Theory. Proceedings, 1978. Edited by M. Denker and K. Jacobs. XII, 209 pages. 1979.

Vol. 730: Functional Differential Equations and Approximation of Fixed Points. Proceedings, 1978. Edited by H.-O. Peitgen and H.-O. Walther. XV, 503 pages. 1979.

Vol. 731: Y. Nakagami and M. Takesaki, Duality for Crossed Products of von Neumann Algebras. IX, 139 pages. 1979.

Vol. 732: Algebraic Geometry. Proceedings, 1978. Edited by K. Lønsted. IV, 658 pages. 1979.

Vol. 733: F. Bloom, Modern Differential Geometric Techniques in the Theory of Continuous Distributions of Dislocations. XII, 206 pages. 1979.

Vol. 734: Ring Theory, Waterloo, 1978. Proceedings, 1978. Edited by D. Handelman and J. Lawrence. XI, 352 pages. 1979.

Vol. 735: B. Aupetit, Propriétés Spectrales des Algèbres de Banach. XII, 192 pages. 1979.

Vol. 736: E. Behrends, M-Structure and the Banach-Stone Theorem. X, 217 pages. 1979.

Vol. 737: Volterra Equations. Proceedings 1978. Edited by S.-O. Londen and O. J. Staffans. VIII, 314 pages. 1979.

Vol. 738: P. E. Conner, Differentiable Periodic Maps. 2nd edition, IV, 181 pages. 1979.

Vol. 739: Analyse Harmonique sur les Groupes de Lie II. Proceedings, 1976–78. Edited by P. Eymard et al. VI, 646 pages. 1979.

Vol. 740: Séminaire d'Algèbre Paul Dubreil. Proceedings, 1977–78. Edited by M.-P. Malliavin. V, 456 pages. 1979.

Vol. 741: Algebraic Topology, Waterloo 1978. Proceedings. Edited by P. Hoffman and V. Snaith. XI, 655 pages. 1979.

Vol. 742: K. Clancey, Seminormal Operators. VII, 125 pages. 1979.

Vol. 743: Romanian-Finnish Seminar on Complex Analysis. Proceedings, 1976. Edited by C. Andreian Cazacu et al. XVI, 713 pages. 1979.

Vol. 744: I. Reiner and K. W. Roggenkamp, Integral Representations. VIII, 275 pages. 1979.

Vol. 745: D. K. Haley, Equational Compactness in Rings. III, 167 pages. 1979.

Vol. 746: P. Hoffman, τ-Rings and Wreath Product Representations. V, 148 pages. 1979.

Vol. 747: Complex Analysis, Joensuu 1978. Proceedings, 1978. Edited by I. Laine, O. Lehto and T. Sorvali. XV, 450 pages. 1979.

Vol. 748: Combinatorial Mathematics VI. Proceedings, 1978. Edited by A. F. Horadam and W. D. Wallis. IX, 206 pages. 1979.

Vol. 749: V. Girault and P.-A. Raviart, Finite Element Approximation of the Navier-Stokes Equations. VII, 200 pages. 1979.

Vol. 750: J. C. Jantzen, Moduln mit einem höchsten Gewicht. III, 195 Seiten. 1979.

Vol. 751: Number Theory, Carbondale 1979. Proceedings. Edited by M. B. Nathanson. V, 342 pages. 1979.

Vol. 752: M. Barr, *-Autonomous Categories. VI, 140 pages. 1979.

Vol. 753: Applications of Sheaves. Proceedings, 1977. Edited by M. Fourman, C. Mulvey and D. Scott. XIV, 779 pages. 1979.

Vol. 754: O. A. Laudal, Formal Moduli of Algebraic Structures. III, 161 pages. 1979.

Vol. 755: Global Analysis. Proceedings, 1978. Edited by M. Grmela and J. E. Marsden. VII, 377 pages. 1979.

Vol. 756: H. O. Cordes, Elliptic Pseudo-Differential Operators – An Abstract Theory. IX, 331 pages. 1979.

Vol. 757: Smoothing Techniques for Curve Estimation. Proceedings, 1979. Edited by Th. Gasser and M. Rosenblatt. V, 245 pages. 1979.

Vol. 758: C. Năstăsescu and F. Van Oystaeyen; Graded and Filtered Rings and Modules. X, 148 pages. 1979.

Vol. 759: R. L. Epstein, Degrees of Unsolvability: Structure and Theory. XIV, 216 pages. 1979.

Vol. 760: H.-O. Georgii, Canonical Gibbs Measures. VIII, 190 pages. 1979.

Vol. 761: K. Johannson, Homotopy Equivalences of 3-Manifolds with Boundaries. 2, 303 pages. 1979.

Vol. 762: D. H. Sattinger, Group Theoretic Methods in Bifurcation Theory. V, 241 pages. 1979.

Vol. 763: Algebraic Topology, Aarhus 1978. Proceedings, 1978. Edited by J. L. Dupont and H. Madsen. VI, 695 pages. 1979.

Vol. 764: B. Srinivasan, Representations of Finite Chevalley Groups. XI, 177 pages. 1979.

Vol. 765: Padé Approximation and its Applications. Proceedings, 1979. Edited by L. Wuytack. VI, 392 pages. 1979.

Vol. 766: T. tom Dieck, Transformation Groups and Representation Theory. VIII, 309 pages. 1979.

Vol. 767: M. Namba, Families of Meromorphic Functions on Compact Riemann Surfaces. XII, 284 pages. 1979.

Vol. 768: R. S. Doran and J. Wichmann, Approximate Identities and Factorization in Banach Modules. X, 305 pages. 1979.

Vol. 769: J. Flum, M. Ziegler, Topological Model Theory. X, 151 pages. 1980.

Vol. 770: Séminaire Bourbaki vol. 1978/79 Exposés 525–542. IV, 341 pages. 1980.

Vol. 771: Approximation Methods for Navier-Stokes Problems. Proceedings, 1979. Edited by R. Rautmann. XVI, 581 pages. 1980.

Vol. 772: J. P. Levine, Algebraic Structure of Knot Modules. XI, 104 pages. 1980.

Vol. 773: Numerical Analysis. Proceedings, 1979. Edited by G. A. Watson. X, 184 pages. 1980.

Vol. 774: R. Azencott, Y. Guivarc'h, R. F. Gundy, Ecole d'Eté de Probabilités de Saint-Flour VIII-1978. Edited by P. L. Hennequin. XIII, 334 pages. 1980.

Vol. 775: Geometric Methods in Mathematical Physics. Proceedings, 1979. Edited by G. Kaiser and J. E. Marsden. VII, 257 pages. 1980.

Vol. 776: B. Gross, Arithmetic on Elliptic Curves with Complex Multiplication. V, 95 pages. 1980.

Vol. 777: Séminaire sur les Singularités des Surfaces. Proceedings, 1976-1977. Edited by M. Demazure, H. Pinkham and B. Teissier. IX, 339 pages. 1980.

Vol. 778: SK1 von Schiefkörpern. Proceedings, 1976. Edited by P. Draxl and M. Kneser. II, 124 pages. 1980.

Vol. 779: Euclidean Harmonic Analysis. Proceedings, 1979. Edited by J. J. Benedetto. III, 177 pages. 1980.

Vol. 780: L. Schwartz, Semi-Martingales sur des Variétés, et Martingales Conformes sur des Variétés Analytiques Complexes. XV, 132 pages. 1980.

Vol. 781: Harmonic Analysis Iraklion 1978. Proceedings 1978. Edited by N. Petridis, S. K. Pichorides and N. Varopoulos. V, 213 pages. 1980.

Vol. 782: Bifurcation and Nonlinear Eigenvalue Problems. Proceedings, 1978. Edited by C. Bardos, J. M. Lasry and M. Schatzman. VIII, 296 pages. 1980.

Vol. 783: A. Dinghas, Wertverteilung meromorpher Funktionen in ein- und mehrfach zusammenhängenden Gebieten. Edited by R. Nevanlinna and C. Andreian Cazacu. XIII, 145 pages. 1980.

Vol. 784: Séminaire de Probabilités XIV. Proceedings, 1978/79. Edited by J. Azéma and M. Yor. VIII, 546 pages. 1980.

Vol. 785: W. M. Schmidt, Diophantine Approximation. X, 299 pages. 1980.

Vol. 786: I. J. Maddox, Infinite Matrices of Operators. V, 122 pages. 1980.

Vol. 787: Potential Theory, Copenhagen 1979. Proceedings, 1979. Edited by C. Berg, G. Forst and B. Fuglede. VIII, 319 pages. 1980.

Vol. 788: Topology Symposium, Siegen 1979. Proceedings, 1979. Edited by U. Koschorke and W. D. Neumann. VIII, 495 pages. 1980.

Vol. 789: J. E. Humphreys, Arithmetic Groups. VII, 158 pages. 1980.

Vol. 790: W. Dicks, Groups, Trees and Projective Modules. IX, 127 pages. 1980.

Vol. 791: K. W. Bauer and S. Ruscheweyh, Differential Operators for Partial Differential Equations and Function Theoretic Applications. V, 258 pages. 1980.

Vol. 792: Geometry and Differential Geometry. Proceedings, 1979. Edited by R. Artzy and I. Vaisman. VI, 443 pages. 1980.

Vol. 793: J. Renault, A Groupoid Approach to C*-Algebras. III, 160 pages. 1980.

Vol. 794: Measure Theory, Oberwolfach 1979. Proceedings 1979. Edited by D. Kölzow. XV, 573 pages. 1980.

Vol. 795: Séminaire d'Algèbre Paul Dubreil et Marie-Paule Malliavin. Proceedings 1979. Edited by M. P. Malliavin. V, 433 pages. 1980.

Vol. 796: C. Constantinescu, Duality in Measure Theory. IV, 197 pages. 1980.

Vol. 797: S. Mäki, The Determination of Units in Real Cyclic Sextic Fields. III, 198 pages. 1980.

Vol. 798: Analytic Functions, Kozubnik 1979. Proceedings. Edited by J. Ławrynowicz. X, 476 pages. 1980.

Vol. 799: Functional Differential Equations and Bifurcation. Proceedings 1979. Edited by A. F. Izé. XXII, 409 pages. 1980.

Vol. 800: M.-F. Vignéras, Arithmétique des Algèbres de Quaternions. VII, 169 pages. 1980.

Vol. 801: K. Floret, Weakly Compact Sets. VII, 123 pages. 1980.

Vol. 802: J. Bair, R. Fourneau, Etude Géometrique des Espaces Vectoriels II. VII, 283 pages. 1980.

Vol. 803: F.-Y. Maeda, Dirichlet Integrals on Harmonic Spaces. X, 180 pages. 1980.

Vol. 804: M. Matsuda, First Order Algebraic Differential Equations. VII, 111 pages. 1980.

Vol. 805: O. Kowalski, Generalized Symmetric Spaces. XII, 187 pages. 1980.

Vol. 806: Burnside Groups. Proceedings, 1977. Edited by J. L. Mennicke. V, 274 pages. 1980.

Vol. 807: Fonctions de Plusieurs Variables Complexes IV. Proceedings, 1979. Edited by F. Norguet. IX, 198 pages. 1980.

Vol. 808: G. Maury et J. Raynaud, Ordres Maximaux au Sens de K. Asano. VIII, 192 pages. 1980.

Vol. 809: I. Gumowski and Ch. Mira, Recurences and Discrete Dynamic Systems. VI, 272 pages. 1980.

Vol. 810: Geometrical Approaches to Differential Equations. Proceedings 1979. Edited by R. Martini. VII, 339 pages. 1980.

Vol. 811: D. Normann, Recursion on the Countable Functionals. VIII, 191 pages. 1980.

Vol. 812: Y. Namikawa, Toroidal Compactification of Siegel Spaces. VIII, 162 pages. 1980.

Vol. 813: A. Campillo, Algebroid Curves in Positive Characteristic. V, 168 pages. 1980.

Vol. 814: Séminaire de Théorie du Potentiel, Paris, No. 5. Proceedings. Edited by F. Hirsch et G. Mokobodzki. IV, 239 pages. 1980.

Vol. 815: P. J. Slodowy, Simple Singularities and Simple Algebraic Groups. XI, 175 pages. 1980.

Vol. 816: L. Stoica, Local Operators and Markov Processes. VIII, 104 pages. 1980.

Vol. 817: L. Gerritzen, M. van der Put, Schottky Groups and Mumford Curves. VIII, 317 pages. 1980.

Vol. 818: S. Montgomery, Fixed Rings of Finite Automorphism Groups of Associative Rings. VII, 126 pages. 1980.

Vol. 819: Global Theory of Dynamical Systems. Proceedings, 1979. Edited by Z. Nitecki and C. Robinson. IX, 499 pages. 1980.

Vol. 820: W. Abikoff, The Real Analytic Theory of Teichmüller Space. VII, 144 pages. 1980.

Vol. 821: Statistique non Paramétrique Asymptotique. Proceedings, 1979. Edited by J.-P. Raoult. VII, 175 pages. 1980.

Vol. 822: Séminaire Pierre Lelong–Henri Skoda, (Analyse) Années 1978/79. Proceedings. Edited by P. Lelong et H. Skoda. VIII, 356 pages, 1980.

Vol. 823: J. Král, Integral Operators in Potential Theory. III, 171 pages. 1980.

Vol. 824: D. Frank Hsu, Cyclic Neofields and Combinatorial Designs. VI, 230 pages. 1980.

Vol. 825: Ring Theory, Antwerp 1980. Proceedings. Edited by F. van Oystaeyen. VII, 209 pages. 1980.

Vol. 826: Ph. G. Ciarlet et P. Rabier, Les Equations de von Kármán. VI, 181 pages. 1980.

Vol. 827: Ordinary and Partial Differential Equations. Proceedings, 1978. Edited by W. N. Everitt. XVI, 271 pages. 1980.

Vol. 828: Probability Theory on Vector Spaces II. Proceedings, 1979. Edited by A. Weron. XIII, 324 pages. 1980.

Vol. 829: Combinatorial Mathematics VII. Proceedings, 1979. Edited by R. W. Robinson et al.. X, 256 pages. 1980.

Vol. 830: J. A. Green, Polynomial Representations of GL_n. VI, 118 pages. 1980.

Vol. 831: Representation Theory I. Proceedings, 1979. Edited by V. Dlab and P. Gabriel. XIV, 373 pages. 1980.

Vol. 832: Representation Theory II. Proceedings, 1979. Edited by V. Dlab and P. Gabriel. XIV, 673 pages. 1980.

Vol. 833: Th. Jeulin, Semi-Martingales et Grossissement d'une Filtration. IX, 142 Seiten. 1980.

Vol. 834: Model Theory of Algebra and Arithmetic. Proceedings, 1979. Edited by L. Pacholski, J. Wierzejewski, and A. J. Wilkie. VI, 410 pages. 1980.

Vol. 835: H. Zieschang, E. Vogt and H.-D. Coldewey, Surfaces and Planar Discontinuous Groups. X, 334 pages. 1980.

Vol. 836: Differential Geometrical Methods in Mathematical Physics. Proceedings, 1979. Edited by P. L. García, A. Pérez-Rendón, and J. M. Souriau. XII, 538 pages. 1980.

Vol. 837: J. Meixner, F. W. Schäfke and G. Wolf, Mathieu Functions and Spheroidal Functions and their Mathematical Foundations Further Studies. VII, 126 pages. 1980.

Vol. 838: Global Differential Geometry and Global Analysis. Proceedings 1979. Edited by D. Ferus et al. XI, 299 pages. 1981.

Vol. 839: Cabal Seminar 77 – 79. Proceedings. Edited by A. S. Kechris, D. A. Martin and Y. N. Moschovakis. V, 274 pages. 1981.

Vol. 840: D. Henry, Geometric Theory of Semilinear Parabolic Equations. IV, 348 pages. 1981.

Vol. 841: A. Haraux, Nonlinear Evolution Equations- Global Behaviour of Solutions. XII, 313 pages. 1981.

Vol. 842: Séminaire Bourbaki vol. 1979/80. Exposés 543–560. IV, 317 pages. 1981.

Vol. 843: Functional Analysis, Holomorphy, and Approximation Theory. Proceedings. Edited by S. Machado. VI, 636 pages. 1981.

Vol. 844: Groupe de Brauer. Proceedings. Edited by M. Kervaire and M. Ojanguren. VII, 274 pages. 1981.

Vol. 845: A. Tannenbaum, Invariance and System Theory: Algebraic and Geometric Aspects. X, 161 pages. 1981.

Vol. 846: Ordinary and Partial Differential Equations, Proceedings. Edited by W. N. Everitt and B. D. Sleeman. XIV, 384 pages. 1981.

Vol. 847: U. Koschorke, Vector Fields and Other Vector Bundle Morphisms – A Singularity Approach. IV, 304 pages. 1981.

Vol. 848: Algebra, Carbondale 1980. Proceedings. Ed. by R. K. Amayo. VI, 298 pages. 1981.

Vol. 849: P. Major, Multiple Wiener-Itô Integrals. VII, 127 pages. 1981.

Vol. 850: Séminaire de Probabilités XV. 1979/80. Avec table générale des exposés de 1966/67 à 1978/79. Edited by J. Azéma and M. Yor. IV, 704 pages. 1981.

Vol. 851: Stochastic Integrals. Proceedings, 1980. Edited by D. Williams. IX, 540 pages. 1981.

Vol. 852: L. Schwartz, Geometry and Probability in Banach Spaces. X, 101 pages. 1981.

Vol. 853: N. Boboc, G. Bucur, A. Cornea, Order and Convexity in Potential Theory: H-Cones. IV, 286 pages. 1981.

Vol. 854: Algebraic K-Theory. Evanston 1980. Proceedings. Edited by E. M. Friedlander and M. R. Stein. V, 517 pages. 1981.

Vol. 855: Semigroups. Proceedings 1978. Edited by H. Jürgensen, M. Petrich and H. J. Weinert. V, 221 pages. 1981.

Vol. 856: R. Lascar, Propagation des Singularités des Solutions d'Equations Pseudo-Différentielles à Caractéristiques de Multiplicités Variables. VIII, 237 pages. 1981.

Vol. 857: M. Miyanishi. Non-complete Algebraic Surfaces. XVIII, 244 pages. 1981.

Vol. 858: E. A. Coddington, H. S. V. de Snoo: Regular Boundary Value Problems Associated with Pairs of Ordinary Differential Expressions. V, 225 pages. 1981.

Vol. 859: Logic Year 1979–80. Proceedings. Edited by M. Lerman, J. Schmerl and R. Soare. VIII, 326 pages. 1981.

Vol. 860: Probability in Banach Spaces III. Proceedings, 1980. Edited by A. Beck. VI, 329 pages. 1981.

Vol. 861: Analytical Methods in Probability Theory. Proceedings 1980. Edited by D. Dugué, E. Lukacs, V. K. Rohatgi. X, 183 pages. 1981.

Vol. 862: Algebraic Geometry. Proceedings 1980. Edited by A. Libgober and P. Wagreich. V, 281 pages. 1981.

Vol. 863: Processus Aléatoires à Deux Indices. Proceedings, 1980. Edited by H. Korezlioglu, G. Mazziotto and J. Szpirglas. V, 274 pages. 1981.

Vol. 864: Complex Analysis and Spectral Theory. Proceedings, 1979/80. Edited by V. P. Havin and N. K. Nikol'skii. VI, 480 pages. 1981.

Vol. 865: R. W. Bruggeman, Fourier Coefficients of Automorphic Forms. III, 201 pages. 1981.

Vol. 866: J.-M. Bismut, Mécanique Aléatoire. XVI, 563 pages. 1981.

Vol. 867: Séminaire d'Algèbre Paul Dubreil et Marie-Paule Malliavin. Proceedings, 1980. Edited by M.-P. Malliavin. V, 476 pages. 1981.

Vol. 868: Surfaces Algébriques. Proceedings 1976–78. Edited by J. Giraud, L. Illusie et M. Raynaud. V, 314 pages. 1981.

Vol. 869: A. V. Zelevinsky, Representations of Finite Classical Groups. IV, 184 pages. 1981.

Vol. 870: Shape Theory and Geometric Topology. Proceedings, 1981. Edited by S. Mardešić and J. Segal. V, 265 pages. 1981.

Vol. 871: Continuous Lattices. Proceedings, 1979. Edited by B. Banaschewski and R.-E. Hoffmann. X, 413 pages. 1981.

Vol. 872: Set Theory and Model Theory. Proceedings, 1979. Edited by R. B. Jensen and A. Prestel. V, 174 pages. 1981.

Vol. 873: Constructive Mathematics, Proceedings, 1980. Edited by F. Richman. VII, 347 pages. 1981.

Vol. 874: Abelian Group Theory. Proceedings, 1981. Edited by R. Göbel and E. Walker. XXI, 447 pages. 1981.

Vol. 875: H. Zieschang, Finite Groups of Mapping Classes of Surfaces. VIII, 340 pages. 1981.

Vol. 876: J. P. Bickel, N. El Karoui and M. Yor. Ecole d'Eté de Probabilités de Saint-Flour IX – 1979. Edited by P. L. Hennequin. XI, 280 pages. 1981.

Vol. 877: J. Erven, B.-J. Falkowski, Low Order Cohomology and Applications. VI, 126 pages. 1981.

Vol. 878: Numerical Solution of Nonlinear Equations. Proceedings, 1980. Edited by E. L. Allgower, K. Glashoff, and H.-O. Peitgen. XIV, 440 pages. 1981.

Vol. 879: V. V. Sazonov, Normal Approximation – Some Recent Advances. VII, 105 pages. 1981.

Vol. 880: Non Commutative Harmonic Analysis and Lie Groups. Proceedings, 1980. Edited by J. Carmona and M. Vergne. IV, 553 pages. 1981.

Vol. 881: R. Lutz, M. Goze, Nonstandard Analysis. XIV, 261 pages. 1981.

Vol. 882: Integral Representations and Applications. Proceedings, 1980. Edited by K. Roggenkamp. XII, 479 pages. 1981.

Vol. 883: Cylindric Set Algebras. By L. Henkin, J. D. Monk, A. Tarski, H. Andréka, and I. Németi. VII, 323 pages. 1981.

Vol. 884: Combinatorial Mathematics VIII. Proceedings, 1980. Edited by K. L. McAvaney. XIII, 359 pages. 1981.

Vol. 885: Combinatorics and Graph Theory. Edited by S. B. Rao. Proceedings, 1980. VII, 500 pages. 1981.

Vol. 886: Fixed Point Theory. Proceedings, 1980. Edited by E. Fadell and G. Fournier. XII, 511 pages. 1981.

Vol. 887: F. van Oystaeyen, A. Verschoren, Non-commutative Algebraic Geometry, VI, 404 pages. 1981.

Vol. 888: Padé Approximation and its Applications. Proceedings, 1980. Edited by M. G. de Bruin and H. van Rossum. VI, 383 pages. 1981.

Vol. 889: J. Bourgain, New Classes of $\mathcal{L}^p$-Spaces. V, 143 pages. 1981.

Vol. 890: Model Theory and Arithmetic. Proceedings, 1979/80. Edited by C. Berline, K. McAloon, and J.-P. Ressayre. VI, 306 pages. 1981.

Vol. 891: Logic Symposia, Hakone 1979, 1980. Proceedings, 1979, 1980. Edited by G. H. Müller, G. Takeuti, and T. Tugué. XI, 394 pages. 1981.

Vol. 892: H. Cajar, Billingsley Dimension in Probability Spaces. III, 106 pages. 1981.